HOW TO MAKE AND SET NETS

By the same Author:

MODERN DEEP SEA TRAWLING GEAR
(first published 1967)

Fishing News (Books) Ltd. are specialist publishers of technical books on deep sea and inshore commercial fishing. An illustrated catalogue of books issued is available free on request. In early 1968 the list included nearly 40 titles and was being steadily increased. New books are regularly announced in our journals 'Fishing News' (weekly) and 'Fishing News International' (monthly) as issued from:

110 FLEET STREET, LONDON EC4

HOW TO MAKE AND SET NETS

or

THE TECHNOLOGY OF NETTING

BY

JOHN GARNER

Published by

FISHING NEWS (BOOKS) LTD.

LUDGATE HOUSE, 110 FLEET STREET, LONDON, E.C.4, ENGLAND

© FISHING NEWS (BOOKS) LIMITED 1962

First published 1962

Reprinted 1968

Reprinted 1974

ISBN 0 85238 031 3

PRINTED IN GREAT BRITAIN
BY THE WHITEFRIARS PRESS LIMITED, LONDON AND TONBRIDGE

CONTENTS

LIST OF DIAGRAMS

APPENDIX

a. an oyster dredge
b. a beam trawl
c. a herring trawl
d. a boom or stow net
e. a gill set net
f. a trammel net

} 75

g. crayfish trap
h. a beach seine
i. corner of a herring drift net
j. a type of lift net
k. hoop net or fyke net
l. shooting drift nets

} 76

m. plan of a trawler's deck
n. detail of trawler's gallows
o. plan of seiner's deck
p. power block purse seine boat
q. great line fishing
r. pair trawling

} 77

LIST OF CHARTS

PUBLISHER'S NOTE

THIS VOLUME is of high value to both the net maker and the practical fisherman who uses and operates those nets.

It is of particular importance in that it recognises the steady swing towards the increased use of machine made nets for certain purposes. By its explanation of the basic technology of preparing the looms in order to make and shape the various nets with maximum efficiency and economy, it will enable the net maker to utilise his equipment to best advantage.

The earlier chapters of the book are, therefore, particularly concerned with these factors and will prove of ready reference value to the machine operator. The later chapters deal with the specific nets used by fishermen and give an understanding of the principles of operation and rigging by which they give best results.

According to an old Chinese proverb, one picture is worth 10,000 words. That principle is applied in this book by the inclusion in the text and appendix of some 60 specific illustrations of nets and gear. These are given in such detail that they illustrate and supplement the written word very admirably.

Of the author's capacity to treat this subject there can be no doubt. Although still a young man he has made netting and fishing gear his hobby as a boy and subsequently his life interest. His father first excited his interest by bringing home as a boyhood toy the model of a trawl net. Thus the skill of catching fish with nets, the designing of nets and the intricacies of shaping and rigging them became the central fascination and enthusiasm of his life. Living in Hull he was able, while still a youngster, to regularly visit the Pickering Museum and study the excellent model nets and fishing boats with their equipment there displayed. This interest was further accentuated during his schooling at the Hull Nautical College in the years 1942–1945.

From school he went to sea and gained practical experience on deep sea craft from 1945–1947. Called up for national service he occupied his leisure in that two year spell in the Navy by braiding all kinds of small model nets and gear ready for his return to fishing.

In 1949 he was again back on trawlers sailing eventually as a mate until in 1953 he took a post with a net manufacturer. This occupation took him to Scotland and at the age of 25 he wrote and illustrated his first book Deep Sea Trawling.

There followed in the same employ a two-year spell at Cape Town and another year in Rhodesia before he returned to Scotland and entered another net factory.

In 1958 he joined a second net manufacturer till June, 1961, when he accepted a position in the Gear Section at the Marine Laboratory, Aberdeen. Before that, however, he had spent two brief spells at Rome as Gear Consultant with the Fisheries Division, F.A.O. This following directly from his wide experience and the authorship of the book mentioned.

In his own words Mr. Garner "has been fortunate in having been able to make a hobby out of nets and netting" and to develop that hobby into pleasurable and engrossing activity in his daily work.

What makes Mr. Garner's work of outstanding value is the fact that he is able by his very competent sketches to convey his knowledge to the reader by such clear and adequate illustrations.

Fishing News (Books) Ltd. is glad to have the opportunity of publishing this work which will prove of real value to both net manufacturers and fishermen.

AUTHOR'S NOTE

This book is dedicated to my wife, for her tremendous help and understanding whilst compiling the information in these pages.

My grateful thanks are also due to a number of others whom I do not need to name because they will know of my appreciation when they read this.

PREFACE
to reprinted edition—1968

THE SUBJECT OF 'Fishing Gear Technology' is vast indeed. Considering its importance, however, it is one which has not, until comparatively recently received the attention it deserves. When this book was first published in 1962 the Preface began: "There are so many different fishing nets and methods of application, as well as variances in terminology from one country to the next, that it would be difficult to include in any one work a complete compilation of this information. An attempt has been made in this book however, to illustrate the basic designs of the principal gears, also to combine and simplify the relevant information so that those interested may be able to understand more fully a subject which is often regarded as complex."

Many others, of course, realised the urgent need for more clarification of fishing gear terminology, and in 1961, when the material contained in the following pages was being completed, the International Organisation for Standardization (IOS) set up a committee of experts to examine the question of standards of terminology for fishing nets, etc. for international usage. A sub-committee—Textile Products for Fishing Nets (SC. 9) of the Technical Committee TC 38 was subsequently formed for this purpose. Late in 1967 this committee was given a status of its own as RCT/8.

The writer was privileged to serve on that sub-committee, and is therefore aware of some of the difficulties which have to be overcome when endeavouring to reach international agreement on so complicated a subject. Certain recommendations have been made however. It is worthwhile noting in advance that whilst these may vary slightly to those initiated in this book, the principles remain quite valid and are likely to prove possibly most useful in correlating the ISO recommendations when they shall eventually be made available for general use.

Although it would be a formidable task to make a complete dictionary of fishing gear and terminology, the one included at the end of this book does give a comprehensive coverage for the purpose of correlating the appropriate text and illustrations contained in the preceding chapters. In this respect many readers have indicated that the dictionary has been helpful for quickly identifying a certain item which might otherwise have been difficult to trace.

Referring once again to the former Preface, it was then said: "Making nets by hand is undoubtedly on the decline, and this book therefore, gives an insight into how nets may be planned and cut from machined sheets of netting."

This statement has certainly proved to be correct; if anything, perhaps the changeover has been quicker than was initially anticipated so that the use of machined netting is now firmly established in all the important fishery-conscious countries. This has happened in spite of some quite strong opposition to the use of machined netting for trawl nets, at the time when this book was first published. Today, in 1968, such products are being used successfully to fish the waters off Africa, Canada, New Zealand and America, as well as in many other seas, including those within the belt of the Arctic Circle. The hand-made trawl in fact, is becoming the exception rather than the rule.

The new reader who may require more information on trawls may find special interest in Chapter 7, particularly the section dealing with Improvement To The Deep Sea Trawl, for although this was prepared in 1959 the suggested methods of improved designing may still be adapted to some trawl net patterns as the ideas offered for modifying the Granton trawl have proved most successful. Of the comments on Mid Water Trawls, appearing later in this chapter, one indicates that the two boat operation is commercially successful, whilst the other points out "it seems likely that in the very near future single boat Mid Water Trawls will be perfected to the extent where they will be in popular use commercially". This is now so, for it is not unusual to see a large European deep sea trawler rigged with both ground and mid water trawls for easy conversion. Inshore boats now also operate mid water trawls in many parts of the world without complications.

With the ever increasing demand on the resources of the seas, it is hoped that this work will continue to be a useful hand book to those in the fishing industries of the advanced and developing countries.

JOHN GARNER.

November 1967.

PREFACE to first edition—1962

There are so many different fishing nets and methods of application, as well as variances in terminology from one country to the next, that it would be difficult indeed to include in any one work, a complete compilation of this information. An attempt has been made in this book, however, to illustrate the basic designs of some of the principal gears, also to combine and simplify the relevant information so that those interested may be able to understand more fully a subject which is often regarded as complex.

One of the reasons for this complexity may be that the fishing industry is so elastic, especially as far as catching is concerned. Consider for instance the conditions which affect fish—climatic, migratory, feeding, tidal, spawning, etc., all of which are changeable, and so consequently are the movements of fish. In view of these variations it does not follow that one net of a specific design which is catching well at a certain period in one area, will be as effective in another area. A slight alteration to the net or rig can make all the difference, and developing from this we have many different net specifications. But really in a final analysis, we are left with only the few basic designs, i.e. the seine net, the trawl, the surround net, the trap net, the gill net and the entangle net.

Making nets by hand is undoubtedly on the decline, and this book, therefore, gives an insight into how nets may be planned and cut from machined sheets of netting.

A dictionary of fishing gear and terminology is included at the end of the book. This should be a useful reference, as most of the terms therein are correlated with the appropriate text or illustrations contained in the pages of the preceding chapters.

CHAPTER I

MESH SIZE

Importance of the mesh and methods of measuring; the 'run' of knots and comparisons in mesh measurement

Measuring Mesh

Reference Fig. 1

THE ALL important factor relevant to any net is the size of the mesh. Besides governing, to some extent, the size of fish that can be caught, it also determines the way in which the netting can be rigged to the ropes.

There are many methods of measuring mesh size, i.e. the total length of the four sides; measuring from the centre of one knot to the centre of the next diagonally opposite knot; the number of rows to a stretched yard; the use of a gauge; or the length of only one side.

Fig. 1 gives an indication of how a mesh would appear when measuring the side. It also shows that one row is half a mesh, as distinct from a bar which is one side of a mesh.

Certain laws covered by the Sea Fishing Industries (Fishing Nets) Order of 1956 govern the minimum size of mesh permissible for different classes of nets, and these call for such dimensions throughout a net when wet that a specified gauge will pass through.

Deep Sea trawl nets used north of the 63rd parallel; every mesh should allow the entry of a 110 mm. (4⅜ in.) gauge.

Small trawls require a mesh of such a size that a 75 mm. (3 in.) gauge will pass through.

Seine Nets have to be made with netting of a mesh size suitable to permit entry of a 70 mm. (2¾ in.) gauge.

Direction of Knots

Reference Fig. 2

THE KNOTS which form the meshes of modern machined netting usually appear as illustrated in Fig. 2 and as with the knots formed when hand braiding, there is a distinct direction with the 'run of the knots' which is discernible from the opposite direction 'against the knots'.

This is an extremely important consideration when rigging, especially where synthetic twines are used, for the 'run' should always be with the direction of the pull or strain which will be apparent when the net is fished.

If one imagines that the meshes shown in the illustration have been cut from a sheet of netting it will be noticed that the loops along the upper edge remain intact even after the broken ends (marked a) have been withdrawn. These meshes, therefore, may be called 'clean meshes'.

Looking at the adjacent edge, it will be clearly seen that the knots (marked b) cannot be untied without completely breaking down the mesh. Consequently the meshes along this edge can be referred to as 'cut meshes'. (*See* Chapter 2: Length and Depth of Netting.)

FIG: I.

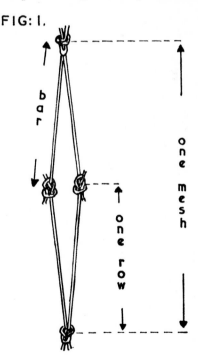

CHART 1. COMPARISONS OF MESH MEASUREMENT

| Size of Mesh (Stretch) | | Rows Per | Rows Per | Rows Per |
Millimetres	Inches	Metre	Alen	Yard
5	·20	400·00	251·00	365·72
10	·39	200·00	125·50	182·86
12	·47	166·66	104·58	152·38
14	·55	142·86	89·64	130·61
16	·63	125·00	78·44	114·29
18	·71	111·11	69·72	101·59
20	·79	100·00	62·75	91·43
22	·87	90·91	57·05	83·12
24	·94	83·33	52·29	76·19
26	1·02	76·92	48·27	70·33
28	1·10	71·43	44·82	65·31
30	1·18	66·66	41·83	60·95
32	1·26	62·50	39·22	57·15
34	1·34	58·82	36·91	53·78
36	1·42	55·55	34·86	50·79
38	1·50	52·63	33·03	48·12
40	1·57	50·00	31·37	45·72
42	1·65	47·62	29·88	43·54
44	1·73	45·45	28·52	41·55
46	1·81	43·48	27·27	39·76
48	1·89	41·66	26·15	38·09
50	1·97	40·00	25·10	36·57
52	2·05	38·46	24·14	35·17
54	2·13	37·04	23·24	33·86
56	2·20	35·71	22·41	32·65
58	2·28	34·48	21·64	31·53
60	2·36	33·33	20·92	30·48
62	2·44	32·26	20·24	29·50
64	2·52	31·25	19·61	28·56
66	2·60	30·30	19·01	27·71
68	2·68	29·41	18·46	26·89
70	2·76	28·57	17·93	26·12
72	2·83	27·78	17·43	25·40
74	2·91	27·03	16·96	24·72
76	2·99	26·32	16·51	24·07
78	3·07	25·64	16·09	23·44
80	3·15	25·00	15·69	22·86
82	3·23	24·39	15·30	22·30
84	3·31	23·81	14·94	21·77
86	3·39	23·25	14·59	21·26
88	3·46	22·73	14·26	20·78
90	3·54	22·22	13·94	20·32
92	3·62	21·74	13·64	19·88
94	3·70	21·28	13·35	19·46
96	3·78	20·83	13·07	19·05
98	3·86	20·41	12·81	18·66
100	3·94	20·00	12·55	18·28

FIG: 2. DIRECTION OF KNOTS.

AGAINST THE 'RUN'.

LENGTH: CLEAN MESHES:

CUT
MESHES:

DEPTH:

WITH
THE
'RUN'

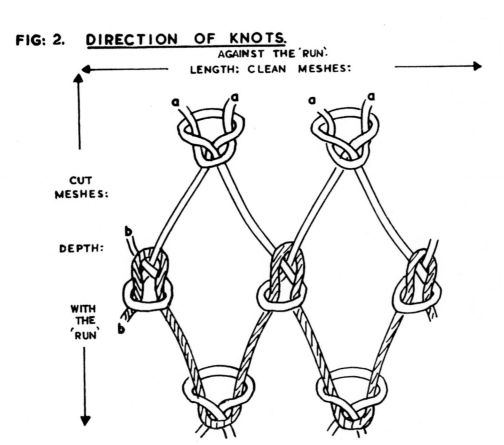

[15]

CHAPTER 2

NET MAKING MACHINES

Three major types of net making machines; methods of adjusting for depth and making knots

Types of Machines

NET MAKING machines may be grouped into three categories, namely single shuttle, multi-shuttle and knotless.

The single shuttle machine, called a 'Mons' or 'Bonamy' loom, was the original instrument for making machined netting, when its working movements were entirely hand operated. In later years powered mechanisation was applied to this type of loom, but it did little to improve a very poor production capacity. Nevertheless, it still has its special uses for making drift nets, ring nets, and certain other nets.

A single shuttle machine forms clean meshes along the width of the machine. The number of meshes made may be between 400 and 720 depending on the kind of machine.

Multi-shuttle machines are numerous in design. They may be made to manufacture single knotted netting with flat knots (the reef knot) or English knots (the sheet bend) or again they may make double knotted netting, with an assortment of variations in the actual knot or bend. Some machines are convertible for manufacturing single or double knotted netting.

A multi-shuttle machine usually forms cut meshes along the width of the machine, and the number may vary between 70 and 500 depending on the gauge of the machine.

The knotless net machine produces meshes which are formed without knots, and this is achieved by running the strands of twine through each other and twisting.

Of the three categories, multi-shuttle net machines are the most widely used, and in the majority of designs the working principle is very similar. To give the reader an elementary insight into this principle it can be classed as the 'Primary Method' of mechanical net making.

The manufacturers of netting machines do issue charts which give all the necessary data relating to capacities, dimensions and production etc. This information gives a guide to the prospective buyer and enables him to select the type of loom which will be most satisfactory for the kind of netting he wishes to have manufactured. The range offered may vary in PITCH[1] from 6 mm. to 15 mm. or even 30 mm. in the case of looms required for braiding heavy trawl twine.

The dimensions of the different makes of looms may vary a little, but to give an idea of the size of the average multi-shuttle design, the approximate dimensions would be: weight $4\frac{1}{4}$ tons, height and width about 7 ft., and length something like 15 ft.

If these measurements are considered it will be obvious that the smaller gauged machine will have more shuttles in the carriage than the heavy gauged machine. Consequently because each shuttle is instrumental in making one mesh it will be readily understandable that a 6 mm. loom with say 403 shuttles and set up to produce netting of 3 in. mesh will be able to turn out a much deeper (width on the loom) sheet of netting, than a 12 mm. machine having only 302 shuttles and making a similar size of mesh. In other words the 6 mm. loom would produce a sheet of netting—in lighter twine of course—wider by one quarter than the netting made by the heavier gauge machine.

Fig. 3 illustrates a netting machine such as we have classed under 'Primary Method'. The material—whether it be cotton, hard fibre, or one of the synthetic range—is fed into the machine by the bobbins banked at the rear of the loom, also by the shuttles carried in the front. Both accessories must be

---

[1] Pitch (or gauge) of a machine is the distance between the shuttle centres.

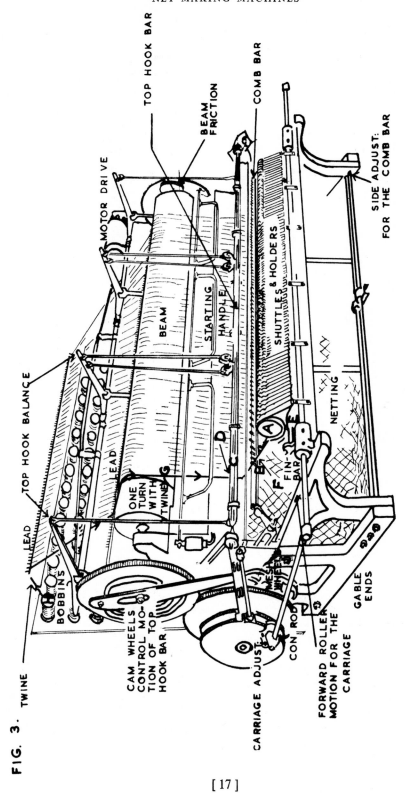

FIG. 3. A NET MAKING MACHINE.

filled with material before the loom is ready for braiding, and very simple winding machines are used for this purpose. The number of bobbins and shuttles depends entirely upon the gauge of the machine, and Chart Number 2 will give the reader an indication of a few of the different sizes and capacities.

The Working Principle of the Primary Method Machine

To FOLLOW a simplified explanation of the working principle (Primary Method) reference should be made to Fig. 4. This drawing is in three parts, which shows the process for forming one knot. It should be remembered though, that in practice, at the time this knot was being made, a net machine would actually be braiding a number of knots which would correspond with the number of shuttles in use on the machine.

A strand of material (G) is shown leading from one bobbin (H), over the beam to meet at the draw (D) another strand of material (F) which comes from shuttle (A).

The appropriate TOP HOOK (C) dips to pick up strand (F) whilst a COMB BAR HOOK (B) raises and grasps the same strand. Lifting, the top hook makes a half turn with strand (F) then moves forward to connect with strand (G) at position 2. Whilst the shuttle (A) is moving, the comb bar hook (B) holds the tension with strand (F) at position 3. Turning downwards the top hook (C) slips the loop made with strand (F) at position 4 and pulls strand (G) through the loop, only to release it again below the shuttle (A) at position 5. The washers (E) which have made a half turn during the above movement resume their original direction, at the same time carrying strand (G) completely under the shuttle holder.

With the concentration of the various tensions the knot is formed at position 6 and as all the working parts move back to their original places, it is drawn over the roller a distance of half the mesh size. Assuming a 3 in. mesh was being manufactured, then the rollers would take the knot 1½ in. away from the draw (6).

The whole process is then repeated, only this time the incliner bar makes a side movement and the knot is formed with the neighbouring strands of twine. Then the incliner bar moves back to its first position, and so on. To allow for this alternating movement, it is necessary to have one bobbing more than there are shuttles; i.e. If a 12 mm. loom is being used to make a sheet of netting 300 meshes deep, 300 shuttles will have to be utilized and 301 bobbins. The completion of the second motion will be netting a mesh in length. Three complete movements will manufacture three rows of netting. Four will equal four rows (or two meshes) and so it continues.

It is often necessary to plan the production of netting in such a way that the machine may be running well below its maximum capacity. That is to say that when making a certain type of net it may be a disadvantage to braid the full width of the machine. For instance; if a sheet of netting 200 meshes deep is required and the machine is a 300 shuttle loom, it may be possible for the machine to be cut down to work efficiently with only 200 shuttles. Naturally this may not sound good economics, but nevertheless it often is.

It does not follow however, that every machine can easily be cut down to produce considerably less meshes than it is designed for. A good deal depends on the type and thickness of the twine which is being braided, also the size of mesh and other factors which can upset the sensitive balance of a machine.

Depth of Netting
Reference Fig. 2

As PREVIOUSLY mentioned, in view of the various methods of fishing and handling different classes of nets, a very important factor in net making is the direction of the knots in relation to a specific sheet of netting. The reason this is so important is because netting is stronger when the pull is 'with the knots'.[1]

With Primary Method machines the knots usually run along the width of the machine,

---

[1] With knotless netting this does not apply.

SHOWING THE MOVEMENT FOR
MAKING ONE KNOT.

FIG: 4

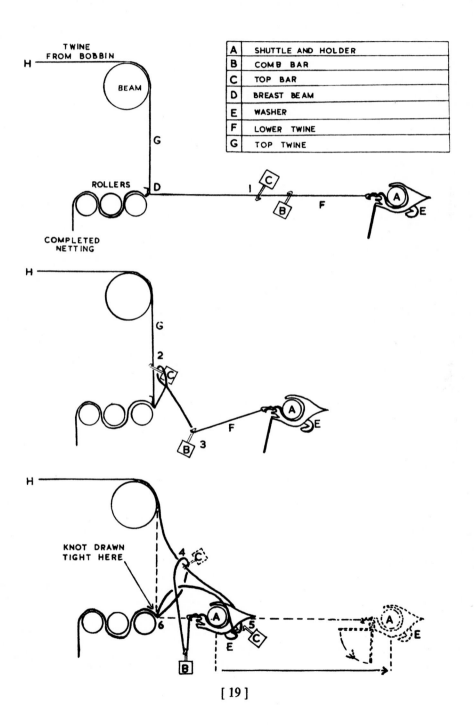

A	SHUTTLE AND HOLDER
B	COMB BAR
C	TOP BAR
D	BREAST BEAM
E	WASHER
F	LOWER TWINE
G	TOP TWINE

TWINE
FROM BOBBIN

BEAM

ROLLERS

COMPLETED
NETTING

KNOT DRAWN
TIGHT HERE

CHART 2. GIVING THE APPROXIMATE CAPABILITIES OF SOME DIFFERENT TYPES OF NET LOOMS

Type of Machine	Pitch in mm.	Number of Shuttles	(Approximate) Weaving Capacities Range of Twine Cotton	Size of Mesh in Inches
Bonamy Loom		400 or 20 Score	36/9 ply to 10/6 ply	$2\frac{1}{4}$ to 4″
Bonamy Loom		720 or 36 Score	40/6 ply to 10/6 ply	$1\frac{3}{4}$ to $2\frac{1}{2}$″
Zang Loom	6	403	40/6 ply to 32/21 ply	$\frac{1}{2}$ to $3\frac{1}{2}$″
Zang Loom	8	403	40/6 ply to 32/21 ply	$\frac{3}{4}$ to $5\frac{1}{2}$″
Zang Loom	10	373	32/6 ply to 32/30 ply	$\frac{3}{4}$ to $7\frac{1}{2}$″
Zang Loom	12	302	32/9 ply to 10/38 ply	$\frac{3}{4}$ to 8″
Porlester Loom	6	404 or 505	100/6 ply to 32/12 ply	$\frac{9}{16}$ to 6″
Porlester Loom	8	404 or 505	40/6 ply to 12/6 ply	$\frac{11}{16}$ to 8″
Porlester Loom	10	404	20/6 ply to 12/15 ply	$\frac{7}{8}$ to 8″
Porlester Loom	13·5	303	32/12 ply to 12/30 ply	$1\frac{3}{16}$ to 8″
Porlester Loom	15	202	12/12 ply to 12/30 ply	$1\frac{5}{16}$ to 9″
Porlester Loom	20	202	12/8 ply to 12/32 ply	$1\frac{11}{16}$ to 10″
Porlester Loom	30	101 or 71	200/3 ply T.T. to 50/4 ply T.T.	$2\frac{3}{4}$ to 16″
Ohls. (Single or double knot)	14	254	40/6 ply to 32/18 ply	$1\frac{1}{2}$ to 9″
German	8	500	40/6 ply to 32/18 ply	$\frac{5}{8}$ to $5\frac{1}{2}$″
Amita (Type 5-4)	5	401	120/6 ply to 20/6 ply	$\frac{3}{8}$ to $7\frac{1}{2}$″
Amita (Type 9-3)	9	301	40/6 ply to 20/15 ply	$\frac{3}{8}$ to $7\frac{1}{2}$″
Amita (Type 14-2)	14	201	20/9 ply to 20/45 ply	$1\frac{1}{4}$ to $7\frac{1}{2}$″
Amita (Type 18-1-5)	18	151	20/12 ply to 20/60 ply	$1\frac{3}{4}$ to $7\frac{1}{2}$″
Amita DKA (Flat single or double knot)		201	20/6 ply to 20s/45	$\frac{3}{4}$ to 18″
Amita DKA (Flat single or double knot)		151	20/12 ply to 20s/75	$1\frac{1}{4}$ to 18″
Amita DKA (Flat single or double knot)		101	20/15 ply to 20/170	$1\frac{3}{4}$ to 18″

and this is referred to in the industry as 'depth'; therefore, the depth of a sheet of netting from a manufacturer's point of view is governed by the number of shuttles in the width of the machine, or to put it another way, the pitch of the machine. *See* Fig. 3.

Length of Netting
Reference Fig. 2
IF THE depth is with the knots then the length must be against the knots; and therefore any feasible length of netting can be manufactured with a 'Primary Method' machine.

[20]

The length of a web of netting is more often than not, referred to by the number of rows, i.e. 800 meshes long would be classed as— length 1,600 rows. Where it is necessary to join two pieces of netting to increase the length, this is referred to as a rough join because the connection is made along the cut meshes.

Supposing, for instance, a sheet of netting was required 800 meshes deep by say, 400 meshes in length, for use in a Purse Seine Net, and that a 'Primary Method' 8 mm. loom with 400 shuttles was to be used for making this; it would be incorrect to plan it on the loom as one sheet 400 meshes by 800 meshes (1,600 rows) for the knots would be in the wrong direction. To be made correctly, it would have to be manufactured in two pieces 400 meshes deep by 400 meshes (800 rows) and then the two identical pieces would have to be joined by hand. The connection would be a clean join because the clean meshes only would be joined together.

Because of the way netting is drawn through the machine during manufacture, the knots may be distorted to some extent, and it is often necessary to stretch it 'with the run' after it has been taken from the machine.

NOTE When planning a net in the width of a loom, the number of shuttles has to be comparable to the number of meshes required. The fact that an additional bobbin or shuttle is sometimes necessary when planning certain types of nets, can mean the forming of an extra half mesh. To avoid confusion, however, all the examples in this book are shown with full meshes.

The figures given on Chart 2 are approximate, and should not be regarded as infallible; for example, the capabilities of the said machines may vary in making single knot netting, as against double knot netting.

CHAPTER 3

LOOM PLANNING

Much economy is possible in making various nets by knowledge of machine capability and using splitting devices

Simple Loom Planning

THERE are certain devices which can be fitted on to a Primary Method net machine, where desired, in such a way, that they continually foul or slip the knot forming process. In doing so, a line of slip knots is made, which can be withdrawn after a sheet of netting has been manufactured, thus leaving two smaller sheets of netting.

For instance, supposing two sheets of netting, each 200 meshes, were required, and a 8 mm. loom was available for manufacturing these, they could be made together, with a splitting device between, i.e. 401 shuttles.

Similarly if a number of light gill nets were required, each 40 meshes deep, using a similar machine with a width of 403 shuttles, the maximum number of nets that could be produced at one time would be nine, accounting for 360 shuttles, plus eight splitting devices.

Another point of interest is the way in which the selvedges can be made on the machine. If, for example, it was desirable to have a half mesh selvedge of double twine along the length (both sides) of each of the gill nets, then the shuttle or bobbin next to the splitting devices would have to be wound with double twine. For a full mesh selvedge two vehicles of double twine would have to be placed at either side of the splitting devices.

Any type of selvedge required along the depth of a net (in this case 40 meshes) has to be made by hand.

Gill nets are usually wanted 'a certain number of meshes deep, by a length of so many yards'. Having described how the depth may be planned we must now consider the second dimension, that of the length, which for manufacturing purposes, must be converted into rows. Imagine, for instance, that the length required is 100 yards, or 3,600 in. of 5 in. mesh, stretched netting, then 1,440 rows would have to be run on the machine. i.e.

$$\frac{3{,}600 \text{ in.} \times 2 \text{ rows}}{5 \text{ in. mesh}}$$

Planning the Machined Manufacture of a Seine Net

Reference Fig. 5

AS EXPLAINED later in the book, the run of the knots has to be lateral in every type of drag net, and therefore, this must be considered when planning the machined manufacture of a seine net.

The sections which have to be made are, the wings; the shoulders; the bag; (this may have to be manufactured in three pieces of differing twine thicknesses) and the cod end. It may be necessary to use different machines for the various sections as determined by factors mentioned earlier, such as the size of the twine, and the dimension of the sections.

Assuming that the specification of the seine net to be manufactured is as follows:—

Wings 100 meshes tapering to 50 meshes in a length of 250 meshes. 5 in. mesh size.

Shoulders 180 meshes tapering to 170 meshes in a length of 100 meshes. 3 in. mesh size.

Bag[1] (1) 400 meshes tapering to 200 meshes in a length of 100 meshes. 3 in. mesh size.

(2) 200 meshes tapering to 100 meshes in a length of 100 meshes. 3 in. mesh size.

Codend 100 meshes straight length 48 meshes. 3 in. mesh size.

The diagrams in Fig. 5 show how the netting for various sections of two similar seines may be produced at one time. The wings, shoulders and bag sections No. 1 may be manufactured on a 302 shuttle loom; whereas the bag sections No. 2 could be produced with a 202 shuttle loom, and the codends

---

[1] In this instance the bag is being planned with two sections only.

FIG: 5.   **PLANNING THE MACHINED MANUFACTURE OF A SEINE NET.**

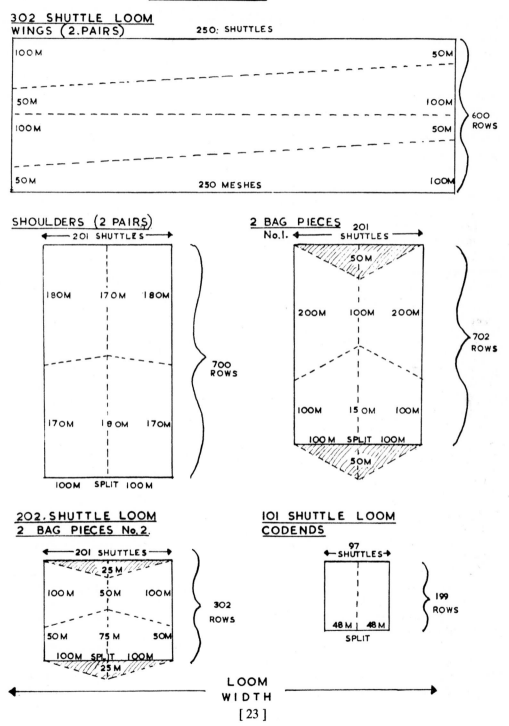

302 SHUTTLE LOOM
WINGS (2.PAIRS)   250: SHUTTLES

100M   50M

50M   100M

100M   50M   600 ROWS

50M   250 MESHES   100M

SHOULDERS (2 PAIRS)
← 201 SHUTTLES →

180M   170M   180M

170M   180M   170M   700 ROWS

100M   SPLIT   100M

2 BAG PIECES
No.1. ← 201 SHUTTLES →

50M

200M   100M   200M

100M   15 0M   100M   702 ROWS

100M   SPLIT   100M

50M

202.SHUTTLE LOOM
2 BAG PIECES No.2.
← 201 SHUTTLES →

25 M

100M   50M   100M

50M   75 M   50M   302 ROWS

100M   SPLIT   100M

25 M

101 SHUTTLE LOOM
CODENDS
← 97 SHUTTLES →

48 M | 48 M   199 ROWS

SPLIT

LOOM WIDTH

[23]

might be made on a 101 shuttle loom. In each case the length becomes the width on the loom.

With the exception of the wings, it will be noticed that there is a splitting device in the middle of each planned sheet.

With the bag net sections and the codends where each cut is to be made there will, of course, be a loss equivalent to a line of half meshes. It is not always essential however, to allow for this when planning, as it is automatically compensated for when the sections are joined together.

With the wings and shoulders, the half mesh lost when cutting is also compensated by the addition of a hand braided selvedge; in fact, to carry planning to its most economic advantage these sections could be planned on the loom with two or three rows less, for the hand made selvedge may be a mesh or more deep.

Planning the Machined Manufacture of a Granton Trawl

When cutting machined sheets of netting into the various sections, it is the usual practice, as explained, to hand selvedge. This is particularly essential where synthetic netting is being cut. With deep sea trawls, made of hard fibre, the position is different, and in view of the life of a deep sea trawl net, it is the considered opinion of the author that, with the exception of the necessary hand made flymesh selvedges along the wings, other hand selvedges are unnecessary. The lacing when the sections are joined together is in fact, adequate protection.

The one main factor against unselvedged sections is that there is not a distinct edge to guide the mender when repair is necessary. This however, might be overcome by running a coloured marker twine through the meshes.

The tremendous asset in favour of the unselvedged edge is the economical aspect, which outshines any minor disadvantages.

Reference Figs. 6 & 15.
Lower wings because of their special shape, are difficult to plan. A Granton lower wing

starts with a base of 50 meshes, which decreases along the inner selvedge (or fishing edge) B diagonally down the 'all bars', the loss being one mesh in two rows. (*See* Fig. 15 Cut A.) The opposite selvedge marked C has to be cut to gain one mesh in three rows. The total loss, therefore, is one mesh in six rows. (*See* Cut B reversed.)

To decrease the lower wing from 50 meshes to 17 meshes the number to be diminished is 33 meshes at the rate of loss i.e. one mesh in six rows (or three meshes). To do this would require a length of 99 meshes, which then becomes the width on the loom.

NOTE The square, top wings and lower wings are all based on a 5½ in. mesh size.

Fig. 6(a) shows how bodies of four lower wings might be planned on say a 101 shuttle loom, in one sheet of netting, 99 meshes × 271 rows. The cut along B is diagonal and top edge D has to be 'rough joined' to lower edge E before the other cuts can be made.

The next stage is to plan the lower wing ends; which remain constant at 17 meshes along the diagonal cut A 'all bars'. It is often the practice at the present time, to omit this part of the trawl net for certain rigs. With the orthodox net, which includes the lower wing ends, a total length of 61½ ft. or 133 meshes has to be accounted for. A length of 99 meshes has already been planned, which on the face of it leaves 34 meshes to be planned. There is, however, (in spite of the half mesh gained when joining) an additional mesh required to make up for the one lost when cutting the diagonals. This means that the length of a wing end should be 35 meshes.

It is the practice when hand braiding, to have several double meshes at the very end of the wing. When running hard fibre twines on net machines, however, it is often difficult to manufacture several double rows at the same time as single, and these usually have to be added by hand.

In order to allow for 10 double rows to be added to each machined lower wing end, the planning need be for only 30 meshes.

This would be as shown in Fig. 6(b). The sheets to be planned on a 71 shuttle loom in two widths, each 30 meshes with the

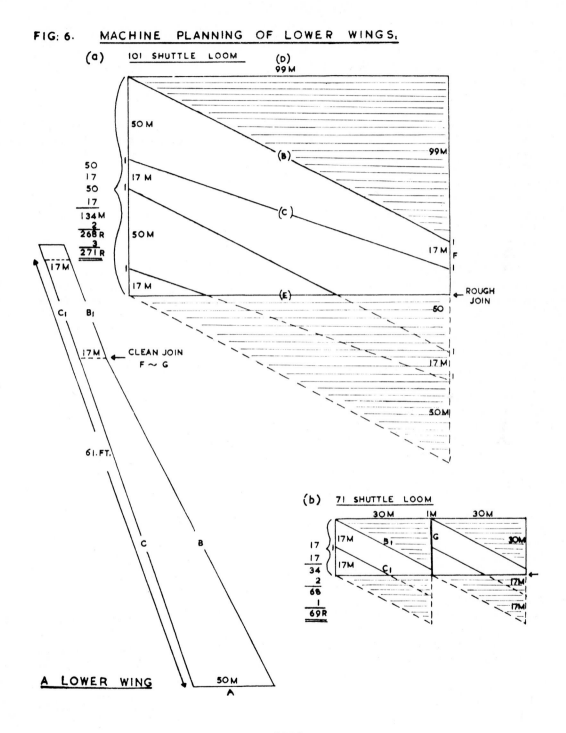

FIG: 6. MACHINE PLANNING OF LOWER WINGS.

(a) 101 SHUTTLE LOOM

A LOWER WING

(b) 71 SHUTTLE LOOM

splitting device between, thus making a web of netting 61 meshes by 69 rows.

IMPORTANT NOTE   A 'rough join' is of course rather unsightly, and not as neat as a 'clean join'. In each example of planning shown, the planned netting has only been sufficient for two similar nets; naturally, however, the more sections that can be planned at one time, the better, for there will be only one 'rough join'.

Supposing for example that the demand was large enough for one specific section, the initial cut might be made and the diagonal section put aside, after which the sections could be continuously manufactured and cut, without having any 'rough joins'.

Once again, when cutting, the diagonals are made, and the top edges are 'rough joined' to the lower edges, after which the four wing end pieces can be cut without wastage.

As will be noted from the example, when the wing end and the main part are 'clean joined' (F to G) double twine is used as a marker row, to facilitate any repairs which may be necessary.

Example

99	meshes	Body of wing
30	meshes	Wing end
5	meshes	Double on
		end

134
less   1½ meshes lost

132½
½ mesh double, for joining
wing end to body
133

Reference Fig. 7

Top Wings   A top wing has a base of 90 meshes decreasing to 11 meshes in a length of 29 ft. 6 in. (or 63/5½ in. meshes). Once again the fishing selvedge, (or flymesh edge) is along the diagonals, but in this case the outer selvedge is straight.

To offset the strain in a hand braided trawl net 16 batings are made in the corners (or quarters) of each top wing. It is not possible to do this on a net machine, and another method must be applied, which is

FIG: 7.   MACHINE PLANNING OF TOP WINGS.

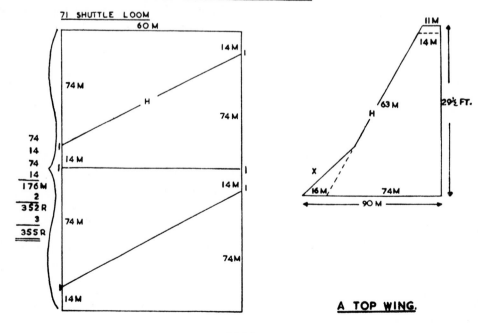

71 SHUTTLE LOOM

A TOP WING.

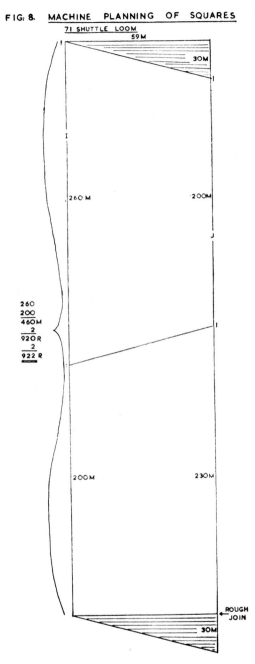

FIG: 8.   MACHINE   PLANNING   OF   SQUARES

71 SHUTTLE LOOM

59 M

30 M

260 M   200M

I

260
200
460M
  2
920R
  2
922 R

200 M   230M

ROUGH
JOIN

30M

90 — 16 = 74 meshes. A small compensating section of netting can be added later, by hand, as shown in the illustration.

Three hand braided meshes of double twine have to be added to the end of the wing, and for planning, these can accordingly be deducted from the length of 63 meshes, leaving 60 meshes to be planned in the width of the loom. 11 meshes are required at the wing end, but in view of the three meshes to be added by hand, which decreases the wing end by three meshes, the planned sections should be for 14 meshes. The required sheet, therefore, is 60 meshes by 355 rows, planned on a 71 shuttle loom.

Reference Fig. 8

The Square Starting at 260 meshes the square decreases to 200 meshes in a length of 60 meshes. There are two rows (one mesh) of double twine at the head of this section, and as the machined squares are planned in reversible positions the double rows must be added by hand after manufacture, therefore, a width of 59 meshes—on, say, a 71 shuttle loom—would be suitable. 922 rows would be sufficient to produce two squares after the triangular piece of netting, which is cut initially, has been 'rough joined' as shown in the illustration.

The planning has, in this case, been shown presuming that hand selvedges were not required. If hand selvedges were desirable, then it would not be necessary to produce quite so many rows.

Reference Fig. 9 & 10

The Bellies When hand braiding, it has been the normal practice in the past, to make the bellies with three panels of netting, each with a different mesh size. The increase in the minimum size of mesh permissible, has really made this practice unnecessary. For this reason, and others, it is far more convenient, without being a disadvantage, to plan the machine made bellies with only two panels.

The bellies are comprised of two similar pieces joined together down the sides. Each complete piece is 200 meshes at the head, and

equally effective. For planning top wings, therefore, the 16 batings are deducted from the number of meshes at the base, i.e.

FIG: 9. MACHINE PLANNING OF BELLY PANELS No. 1.

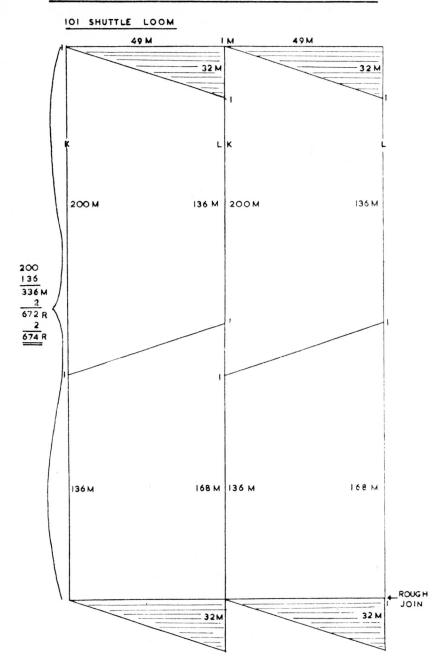

101 SHUTTLE LOOM

FIG: 10. MACHINE PLANNING OF BELLY PANELS No 2.

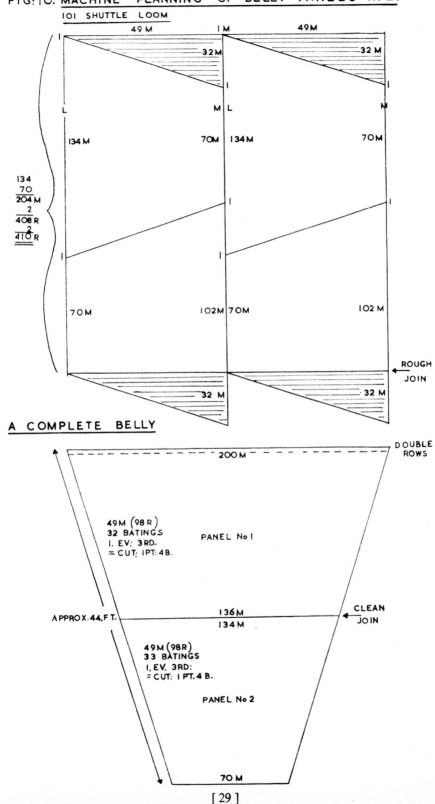

101 SHUTTLE LOOM

A COMPLETE BELLY

decreases down to between 60 and 80 meshes in a length of 44 ft.

Figs. 9 & 10 show how the netting for the two different panels of two pairs of bellies might be planned on a 101 shuttle loom. i.e. One sheet 99 meshes × 674 rows × 5½ in. mesh size, and the other 99 meshes × 410 rows 5¼ or 5 in. mesh size.

Each belly is required 200 meshes tapering to 70 meshes in a length of approximately 44 ft. The rate of loss is standard along both selvedges at one mesh for every three rows, including one lost at each side of the 'clean join', when the panels are connected.

Reference Fig. 11
The Codend As far as planning is concerned, the codend sections are quite straight forward. Using a 71 shuttle loom a width of 68/5¼ in. meshes would give a length of approximately 30 ft. when made. When two of the similar sections are laced together to make top and bottom, the total number of meshes would be 140 round.

The completed Granton trawl is shown in Fig. 12 on page 31. For a possible improved type see page 53.

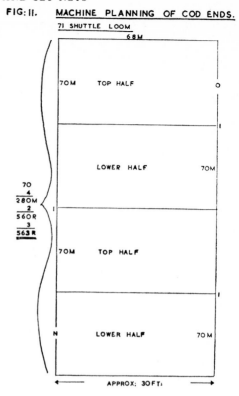

FIG: II.　**MACHINE PLANNING OF COD ENDS.**

71 SHUTTLE LOOM

68 M

70 M　TOP HALF　o

LOWER HALF　70 M

$\frac{70}{4}$
$\overline{280M}$
$\frac{2}{560R}$
$\frac{3}{563R}$

70 M　TOP HALF

N　LOWER HALF　70 M

←　APPROX: 30 FT.　→

A GRANTON TRAWL

FIG: 12

CODLINE

COD END

O

M N

PANEL No 2

BELLYLINE

BELLIES

L

PANEL No I

SQUARE

J K

A

BUNT

B

LOWER WING

C

TOP WING

X

QUARTER

H

SHOWING HALF TOP SIDE

SHOWING HALF LOWER SIDE

B₁

C₁

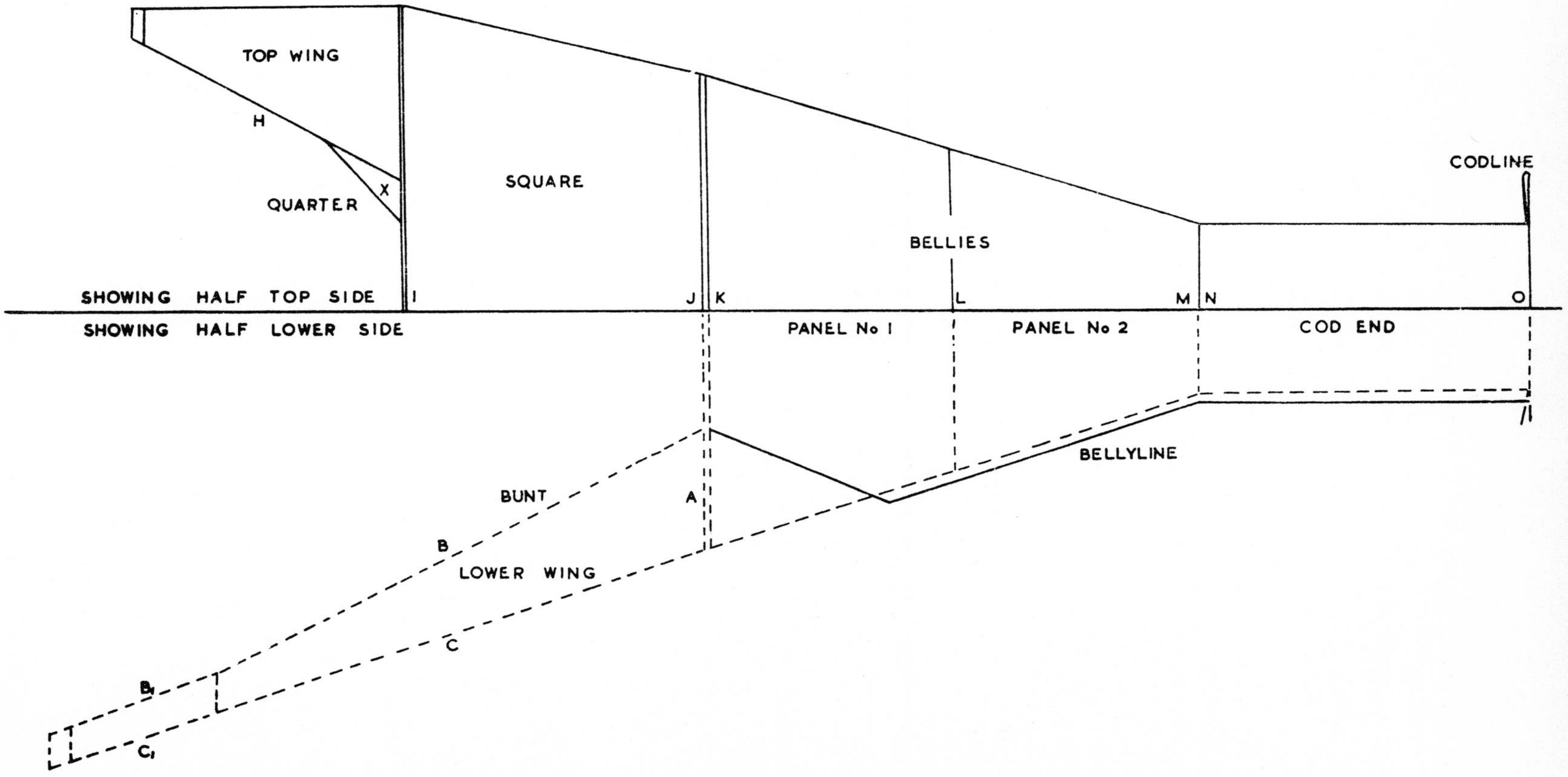

CHAPTER 4

SHAPING NETTING

Extra tapering facilities are available with machine made netting for drag nets, seines, and trawls which require shaping

Tapering and Cutting

Reference Fig. 13, 14 & 15

THERE ARE only two ways of tapering when hand braiding netting, namely: with batings or flymeshes, whereas with machine made netting, which invariably has to be cut to shape, the degree of taper obtainable is almost unlimited.

For cutting machined netting the terms 'rows' and 'meshes' often reverse their positions to those when planning, and the width on the loom becomes—'a number of rows'—while the length is classed as—'so many meshes'—. This may seem unnecessary, but there is a reason for the conversion, which will be apparent if one considers that the nets which most frequently have to be cut to shape, are the various types of drag-net, such as seines and trawls. The strain on these nets is lateral, and the 'run' of the knots must, therefore, be between the wings and the tail of the net—which is of course precisely the direction requiring the shaping.

To understand this more fully, examine Fig. 13 as illustrated. This shows an elevation view of a seine net. Looking at the bag it will be seen that both edges are shown sloping into the tail, and to achieve such taper with a superior cut, it is naturally far better to work with half meshes (or rows) than to use full meshes. This then is the reason for the change.

To make the correct cut one must be acquainted with the terminology which is simplified in Fig. 14. From this, it will be noted that the two sides of a mesh make what is known as a 'point' and the one side of the next mesh which runs in line is called a 'bar'. The cut combining the two is referred to as one point one bar, which would decrease the side of a net by one mesh in every six rows of length. The upper sides of a mesh (clean side) are called a 'mesh', but again the side of the next mesh running in line is known as a 'bar'. The cut shown along the top edge therefore, is one mesh one bar.

There are inferior methods of tapering netting, such as cutting in a full mesh as required.

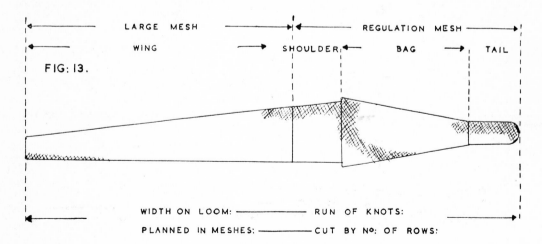

LARGE MESH   REGULATION MESH

WING   SHOULDER   BAG   TAIL

FIG: 13.

WIDTH ON LOOM: —————— RUN OF KNOTS:

PLANNED IN MESHES: —————— CUT BY Nº: OF ROWS:

That is to say, if the desired rate of loss is to be one mesh in six rows, the cut could be made by taking a full mesh for every three meshes of the length, and so on.

FIG: 14.

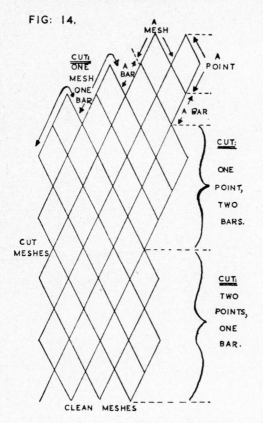

The cutting rate to decrease one mesh in five rows is three points four bars. In practice, however, this would make an uneven taper, and an improved cut giving a similar result is

one point two bars } three points four bars
two points two bars }

A few of the principal cuts are listed below. Some of these are illustrated in Fig. 15 which shows a sheet of netting as it would appear hanging horizontally to half of the stretch measure. The angles of degrees at the foot of the Fig. are the approximate angles of taper for the various cuts which are shown. Naturally if the netting was hung differently, say by one third, then the angles would change accordingly.

	CUT	LOSS
A	All bars	1 mesh in 2 rows
B	1 pt. 4 bars	1 mesh in 3 rows
C	1 pt. 2 bars	1 mesh in 4 rows
D	1 pt. 1 bar × 2 } 1 pt. 2 bars }	1 mesh in 5 rows
E	1 pt. 1 bar	1 mesh in 6 rows
F	1 pt. 1 bar × 3 } 2 pts. 1 bar }	1 mesh in 7 rows
G	1 pt. 1 bar } 2 pts. 1 bar }	1 mesh in 8 rows
H	2 pts. 1 bar × 3 } 1 pt. 1 bar }	1 mesh in 9 rows
I	2 pts. 1 bar	1 mesh in 10 rows
K	3 pts. 1 bar } 2 pts. 1 bar }	1 mesh in 12 rows
L	5 pts. 1 bar } 6 pts. 1 bar }	1 mesh in 24 rows
M	All points	None
N	1 mesh 1 bar	1 mesh in 3 mesh
O	1 mesh 2 bars	1 mesh in 2 mesh

The above results can be checked by referring to Fig. 15. The numbers along the bottom indicate meshes, whilst those down the side represent the number of rows.

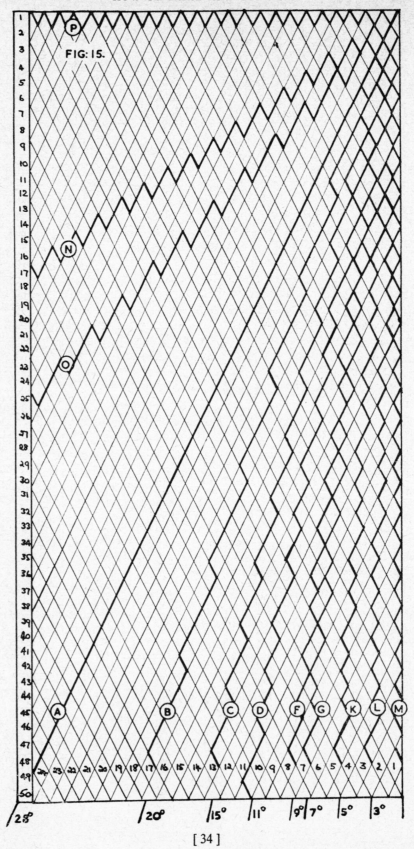

FIG: 15.

CHAPTER 5

HANGING NETTING

Hanging a net to its ropes to secure correct distribution or degree of looseness is illustrated by five examples on gill, surround, drag nets and trawls, with various hitches

Ratios of a Mesh to Different Hanging Proportionates

Reference Fig. 16

THE QUESTION of hanging the netting or a net to ropes with the correct distribution or percentage of looseness is a ticklish one. In this respect, Fig 16 should be studied very carefully. This shows five examples of the formation of a mesh when hung to different proportionates. The values shown are: A. The horizontal mesh opening, which is the percentage figure of the stretched mesh; B. shows the vertical lift of the mesh, and C. shows the height or the vertical opening of the mesh.

Example V. Indicates the appearance of a fully stretched mesh with 100 per cent. closure.

Example W. Shows the appearance of a mesh when hanging in 75 per cent. of its stretched length. The result is a vertical opening of 97 per cent. and a lift of only 3 per cent.

Example X. Illustrates a mesh which has been hung by a half, or to 50 per cent. of its stretched length. In this case the vertical opening is 87 per cent. (It will be noted that this is only 10 per cent. less than the lift for a mesh hung to 25 per cent. of its stretched length.) The upward lift is 13 per cent.

Example Y. Describes a mesh when hung in by one third i.e. 33⅓ of its stretched length. The vertical opening is 74 per cent., giving a lift of 26 per cent. This is undoubtedly the most widely used hanging ratio.

Example Z. This has been illustrated to show the relative percentages when a mesh is hung apparently square. It is identified with the term 'by a third, plus'. The horizontal and vertical openings each show 71 per cent.

and the lift 29 per cent. To hang netting so that the meshes form perfect squares the percentage of looseness to be allowed is 29 per cent.

Hanging Ratios: Netting to Ropes

As with all things relating to nets and netting, hanging procedure has its share of traditionalism, with one group of men preferring one method, and another group a slightly different variation. Again it would be a tremendous task to cover all the minor diversities, for while some minor variations from the standard may be effective, the majority of these differences are so slight as to be of no real consequence.

Hanging Gill Nets

The most straightforward mode of hanging is probably that of attaching gill netting to the ropes. A gill net is frequently hung with a percentage of between 'a half' and 'a third' (*See* Fig. 17 *Examples 1, 2 & 3*). Supposing then that a straight sheet of netting, 40 meshes deep by a length of 100 yd. with a 6 in. mesh, was required, hung 'by a half', then the correct measurement along the frame ropes would be 50 yd. at the top and bottom of the net, and the side ropes would be 87 per cent. of 20 ft. (i.e. 40 m. × 6 in. mesh) = 17·4 ft. To hang a similar sheet 'by a third' the length of the top and bottom ropes of the net would be 66·6 yd. and the side ropes would be 74 per cent. of 20 ft. = 14·8 ft.

Hanging Surround Nets

For mounting beach seine nets and other types of surround nets it is often necessary, in order to effect bagging or pouching, to attach the different sections of netting in varying degrees of looseness. If for instance

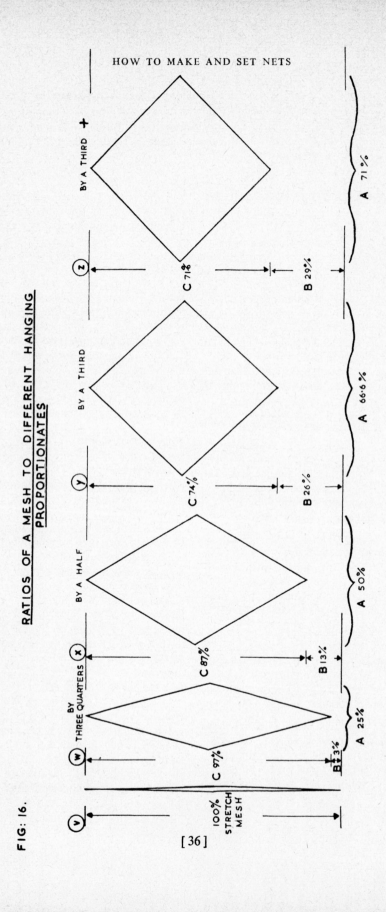

FIG: 16.

RATIOS OF A MESH TO DIFFERENT HANGING PROPORTIONATES

a beach seine net was required in three sections; a centre bag and two outer wings. Although the three sections might be of a similar depth, the bagging can be achieved by hanging the centre bag 'in by a half' and the wings 'in by a third' or even tighter, thus, the wing netting is stretched more along the ropes than the bag, so that when operated the centre of the net automatically pouches. A similar effect can be achieved by making the wings of lesser dimensions in depth than the bag. The three sections are then joined to an equal depth with intakes (or batings) which naturally effect bagging of the larger centre section even though the entire netting, top and bottom is mounted with a similar percentage of looseness. The scooping effect may be enhanced by mounting the netting slightly tighter to the bottom rope.

Hanging Drag Nets

For hanging the more complicated drag nets, such as seines or trawls, the procedure is more involved with a number of important factors to be accounted for. The fundamentals of any drag net, regardless of type, are the frame ropes which connect to the towing ropes, meaning the headline and footrope. The netting is secondary and should be designed to fit the curvature which the ropes will describe when the net is towed. Of course, design and rig of a drag net is more involved than this, with filtering coefficient, drag, and the species to be caught, having to be considered.

As explained, the more stretched the netting is when hung to the ropes, the greater the lift, or the shorter the depth will be. With a seine net it is essential to have two long narrow wings, and the usual method is to hang the wings tight with only 10 per cent. of its stretched length, that is to say, that the length of the main ropes would be 90 per cent. of the length of the stretched netting. This of course still gives a mesh opening of almost 50 per cent. in the depth of the wing. The shoulders of a seine net may be hung in with a looseness of 20 per cent. of the stretched length, giving a mesh opening depth of 80 per cent. Bosoms or centres of a seine

net are usually hung by a half, or even less.

As already intimated there are many diverse variations in hanging, and one common practice with a seine net, is to gain overhang by mounting the netting a fraction looser on the headrope than on the footrope, thus drawing the head of the net into a slightly forward position.

Hanging trawl nets to ropes is again even more complex, for the formation of the netting is entirely different. The wings of a trawl are formed along the diagonal, invariably with flymeshes, and the percentage of looseness to be allowed has to be looked at from another angle. The flymesh side of the top wing which joins to the headline does not call for any real looseness, but should be planned and hung to the ropes with a minimum of slack netting, which should be sufficient to eliminate any strain, and in no way hinder the vertical opening or spread. With the lower wings, which are usually as much as one sixth longer than the total length of the square and top wings, the position is different, and they are usually mounted with a looseness of about 18 per cent. of the stretched length, which, in the bottom trawl, effects the necessary bellying. The lower bosom is usually mounted by the half.

With a large drag net it is often necessary to strengthen the netting with supporting ropes, and these are usually joined from some focal point of strain, such as the quarters where the wings meet the square or belly, from whence they are laced along the diagonal bar in such a way that any stress is distributed along the rope, and not the netting.

Attaching Netting to Ropes
Reference Fig. 17 & 18

There are a number of varied hitches which can be used when connecting netting to ropes, and each different hitch has quite often a distinct application when it can be used to advantage. For instance, the labour involved when making a rolling hitch is greater than when making a clove hitch, but the former hitch has the advantage of being more binding and, therefore, should be used on hard ropes. To save time a clove hitch may be

FIG: 17.
ATTACHING NETTING TO ROPES

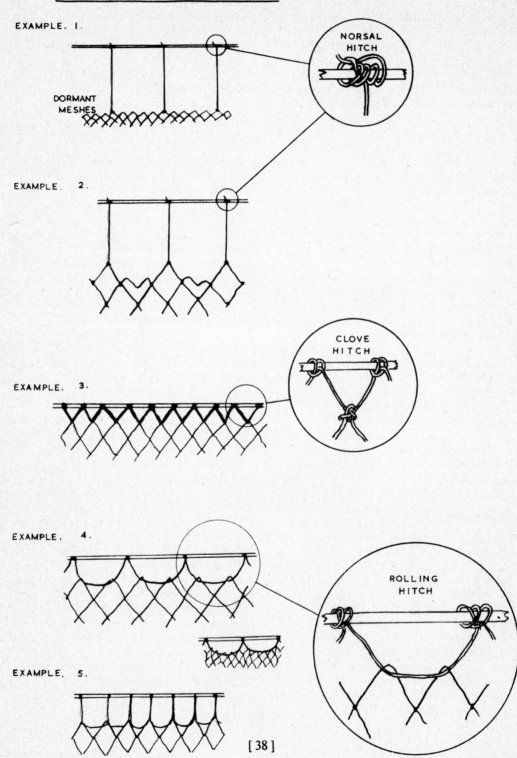

EXAMPLE. 1.

NORSAL HITCH

DORMANT MESHES

EXAMPLE. 2.

CLOVE HITCH

EXAMPLE. 3.

EXAMPLE. 4.

ROLLING HITCH

EXAMPLE. 5.

[38]

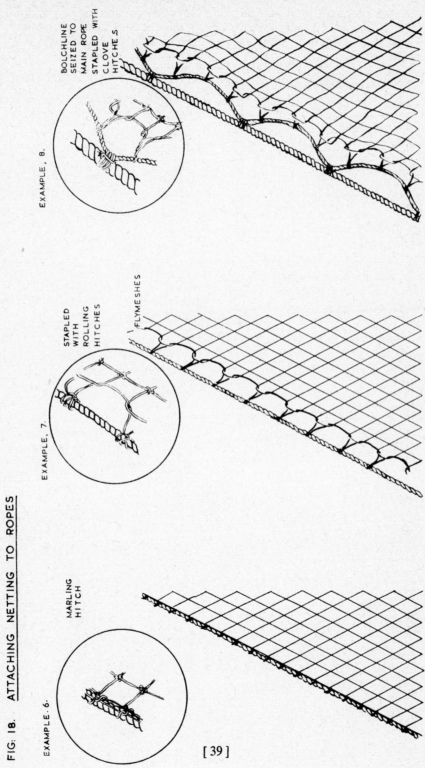

FIG: 18. ATTACHING NETTING TO ROPES

EXAMPLE, 8.

BOLCHLINE
SEIZED TO
MAIN ROPE
STAPLED WITH
CLOVE
HITCHES

EXAMPLE, 7.

STAPLED
WITH
ROLLING
HITCHES

FLYMESHES

EXAMPLE. 6.

MARLING
HITCH

[39]

used on soft ropes such as balch line where the grip will be adequate.

Where netting is joined to ropes via norsals, which are specially laid strands of twine, the hitch used is designed to bind itself so that it does not slip one way or the other. Norsals are threaded through the meshes at intervals and attached to the ropes. The measurement between the netting and the ropes may vary in accordance with the type of net, and it seems that the longer the norsal the more sensitive the net. Naturally the spacing of the norsals is to some extent governed by the size of the mesh.

The following are a few examples of some of the methods used for attaching netting to ropes (see designs, fig. 17 and 18):

Example 1. Shows how a herring drift net is hung with the norsals spaced every four or five meshes; the meshes having been previously strengthened with a cord run along the edge.

Example 2. Illustrates larger mesh netting such as a salmon drift net which may be norsalled every alternate mesh.

Example 3. The method of connecting shown here may be used for attaching trap netting (a Salmon Bag Net or Pound Net) to the ropes, or for hanging the bosoms of small drag nets etc. The connection is made from one mesh to the rope to the next mesh and back to the rope; the joining twine is hitched to the rope and knotted to the mesh.

Example 4. This method may be used for attaching seine net wings or shoulders to the ropes. It is called stapling. Small mesh netting such as ring nets may also be stapled lifting four or more meshes with each bite.

Example 5. This illustrates another variation of stapling.

Example 6. Shows how the wings of a small trawl might be marled to the working ropes.

Example 7. This illustration shows how the wings of a larger trawl might be stapled to the working ropes.

Example 8. Gives an illustration of attaching the wings of a deep sea trawl to the ropes, via the balch line.

CHAPTER 6

SEINE NET PATTERNS

The popular demersal seine net adjusted to white and/or plaice fishing, with detailed specifications for parts and rigging

Seine Nets

Reference Fig. 19, 20 & 21

SO MANY different net designs are referred to as 'seines' that it leads one to think almost any net which is between a gill net and a trawl can be called a 'seine' of one sort or another. i.e. purse seine, beach seine, half-round seine, haddock seine, plaice seine, etc. Each of the nets which have been mentioned is different and such reference can, therefore, be confusing to the layman. The dictionary at the back of the book should throw some light on this.

The nets being dealt with under the above heading are demersal seine nets, that is nets which are dragged over the seabed, in a completely different fashion to trawl nets, and without otter boards. The average power range of the boats may vary considerably and naturally different net specifications are made to suit the different classes of craft.

In Britain the range covers boats of between 40 and 140 horse power. Although nets are made of different dimensions the design remains fairly stereotyped and Fig. 19 illustrates a typical white fish seine. For catching plaice a somewhat different design is more suitable, although the basic principle does not alter, and Fig. 20 shows a drawing of a plaice seine such as may be used by a boat of average power.

The method of seine net fishing is a very effective and reasonably economical way of fishing, and regardless of the class of boat the pattern of working is unchanged. After the direction of the tide has been ascertained one end of rope is moored and the vessel begins to steam in an arc paying out the appropriate number of coils of rope. It may be seven 120 fathom coils, or as many as twelve or thirteen coils, depending on conditions, such as the kind of seabed. The more

rope there is laid out on the bottom the larger the area fished will be. The net is then released and an equal number of coils of rope are payed over the vessel's side as she completes a triangular course back to the mooring, where the first coil of rope is retrieved.

Whilst the boat steams slowly ahead the two ropes are hauled gradually so that they act as 'ticklers' and frighten the fish inwards towards the path of the net. The ropes are re-coiled during the winch hauling operation. Once the net appears it is pulled up and a becket is fixed around the codend so that the bag of fish can be heaved inboard.

Another form of seining is to anchor whilst hauling the gear.

Seine Net Specifications

Reference Fig. 19 (Also 5 & 13)

The Bag. The size of a seine net is usually indicated by the number of meshes round the wide end of the bag, (marked with the arrowed dotted line from position Q). The number may vary between 300[1] meshes and 600 meshes round. The length Q to H also varies from 6 fathoms to $8\frac{1}{2}$ fathoms, tapering to from 80 meshes up to 120 meshes, (round dotted line G to H). The bag is made up of two half sections, each comprising three different thicknesses of netting, the heaviest being at the narrowest end. The two pieces are joined along the length at the top and the bottom.

The Codend or Tail. As far as is possible, the codend has to match up with the number of meshes at the tail-end of the bag (G to H). It is a straight funnel of netting of a length between 9 ft. and 15 ft.

---

[1] A plaice seine net may be only 200 meshes round the bag.

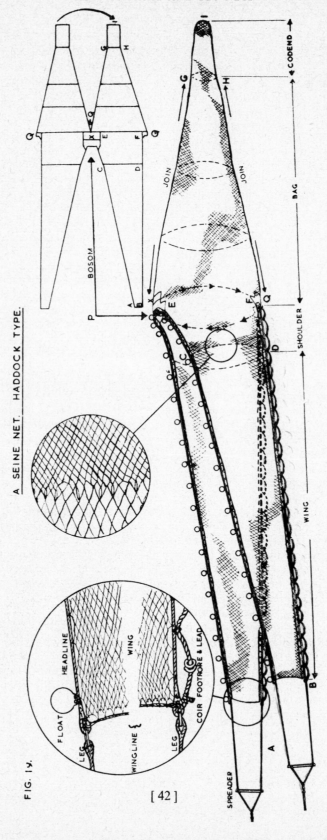

A SEINE NET. HADDOCK TYPE.

FIG. 19.

[42]

FIG: 20. **THE PATTERN OF A PLAICE SEINE**

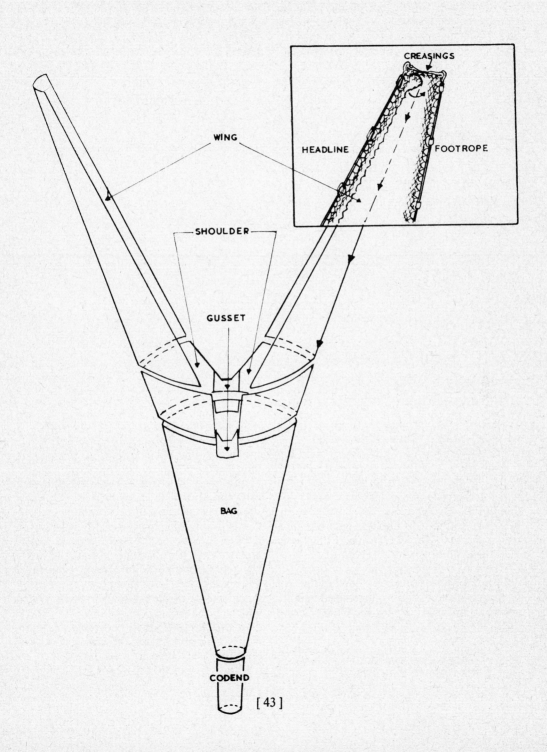

CREASINGS

WING

HEADLINE   FOOTROPE

SHOULDER

GUSSET

BAG

CODEND

[43]

The Shoulders and the Wings.[1] The shoulders and wings are, more often than not, cut with a taper on the headline side, and straight on the lower side. The length of the shoulder (D to Q) may be as little as 10 ft. or as much as 28 ft., while the number of meshes at the wide end (E to F) may be between 280 meshes and 135 meshes. The number of meshes lost in a shoulder is between 10 and 20 making the

inserted as necessary. The fitting of the shoulders and hand-made gussets is, however, more difficult to follow. In some cases the two gussets, which are the bosom strengthening pieces, may be joined in level at the top and bottom of the net, although the usual preference is to have a small piece of netting (x) fitted to the upper edge of the mouth before the top gusset is joined into position. This

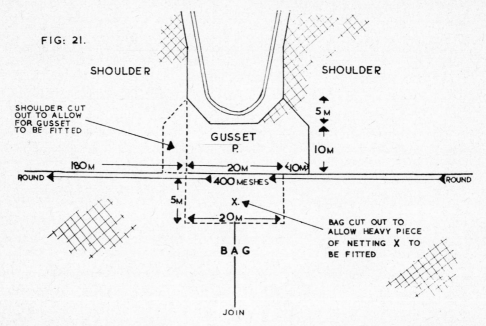

FIG: 21.

average number of meshes (from C to D) anywhere between 260 and 115 meshes.

The wings being of a larger mesh size than the shoulders, possibly twice the size (4½ in. to 5½ in.) do not require as many meshes to make the dimension reasonably equal for joining to the shoulder, the range being between 70 and 100 meshes, tapering in a length of from 12 to 16 fathoms, to anywhere from 30 to 50 meshes (along A to B).

Joining the Netting Sections. Joining the bag sections together down to the codend is quite simple to understand, with the batings being

gives the head of the net a slight overhang or 'sharks mouth'. This overhang can be further enhanced when hanging. Fig. 21 shows how the gussets may be fitted.

Assuming the circumference round the wide end of the bag is 400 meshes, then each shoulder would have to be 180 meshes where it was to join the bag, which would leave 20 meshes for the top bosom and an equal number for the bottom bosom. If each shoulder tapered to 170 meshes in a length of 100 meshes, the loss along the upper edge would be 10 meshes. However, as it is not wise to begin to taper the shoulder until well clear of the gusset, the first decreasing mesh would be made at say the 20th mesh, and after that, one every 8th.

If the wider end of the wing was 100 meshes,

---

[1]In isolated nets the shoulder netting may be cut straight, or with a taper along the centre of the wings as in the case of some plaice seine nets.

the bating ratio for joining to the narrow end of the shoulder, which is 170 meshes, would be 10 to 17, and therefore, 17 shoulder meshes would have to be taken up for every 10 wing meshes. (*See ringed drawing.*) Now supposing the wing decreased to 40 meshes in a length of 200 meshes the loss would be one mesh for every fifth along the upper edge.

The bag would taper in accordance with the length, the baiting or stollers being 'taken-up' equally along the joins.

Rigging. The headline and footrope may be of $1\frac{1}{4}$ in. circumference combination rope, with the headline several feet shorter than the footrope. This is achieved, first by the fact that the top gusset is more forward than the lower one, and second, by attaching the top edge of the shoulder netting tighter than the lower edge along the ropes. For instance, the standard method of mounting the shoulders is to hang two meshes in a bite equal to the length of $1\frac{1}{2}$ stretched meshes. In the case of a 70 mm. mesh (approx $2\frac{3}{4}$ in. mesh) the bite would be about 4 in. therefore, if the shoulder netting was to be hung, on to the top rope with 4 in. bite and on to the bottom rope with $4\frac{1}{4}$ in. bite, the overhang would be increased by more than 2 ft. (100 meshes × $\frac{1}{4}$ in. = 25 in.) along each shoulder.

The wing netting could then be mounted equally on the headline and footrope, with bites equal to $1\frac{1}{2}$ stretched wing meshes.

Floatation can of course vary between 30 to 50—5 in. spherical floats. A 3 in. coir rope is usually attached in bites to the footrope as a protection against chafe, and the lead ring can be threaded along this rope. 150 to 250—$\frac{1}{4}$ lb. ring leads may be used for weighting.

CHAPTER 7

TRAWLS

Large and small trawls described. Improvements to the Granton trawl could increase 'mouth-lift' 7 per cent. and 'spread' 12 per cent.

Otter Trawls

Reference Fig. 22, 23, 24 & 25

THE ADVENT of otter boards must have been a tremendous improvement upon the old beam trawl principle. Later the introduction of the Vigneron Dahl gear and then the improved assembly must certainly have been instrumental in establishing the otter trawl as one of the most successful and widely practised modes of fishing in Europe. Fig. 22 shows the fundamental sections of netting which make a trawl net. These sections may vary in mesh size, taper and dimensions to suit the size and power of the boat which will tow the net. There are hundreds of trawl net specifications, but regardless of size they are basically similar, and composed of the sections shown. The following specifications are two examples:

A Large Trawl

Square: 300 meshes down to 200 meshes. Length 40 ft.

Top wings: 100 meshes down to 12 meshes. Length 46 ft.

Lower wings: 200 meshes down to 34 meshes. Length 96 ft.

Belly (2): 200 meshes down to 100 meshes. Length 24 ft.

Cod end: 100 meshes down to 50 meshes. Length 24 ft.

Mesh size: 5/5½ in.

Headline: length 110 ft.

Groundrope: 68 ft. × 20 ft. × 68 ft.

A Small Trawl

Square: 200 meshes down to 150 meshes. Length 11 ft.

Top wings: 66 meshes down to 11 meshes. Length 21 ft.

Lower wings: 36 meshes down to 20 meshes. Length 35 ft.

Belly (2): 150 meshes down to 60 meshes. Length 35 ft.

Cod end: 60 meshes to 60 meshes. Length 15 ft.

Mesh size: 3/5 in.

Headline: length 60 ft.

Groundrope: 27 ft. × 14 ft. × 27 ft.

Reference Fig. 23 & 24

The gear attachments of a trawl also differ in accordance with the size of the actual net specification and the type of ground to be fished. Fig. 23 shows the Vigneron Dahl type of working gear as used in conjunction with the deep sea trawl.

Looking at Fig. 22 and 23 it will be seen how the working gear fits to the netting to make a complete working trawl. As intimated, however, there are variations from this gear assembly even for a deep sea trawl.

In some circumstances it may be beneficial to use a heavy ground rope rather than bobbins. For example on a smooth sandy bottom where the netting can be allowed to drag in close proximity to the seabed without any real danger of becoming fouled. Such a ground rope may be made of a heavy centre steel wire rope which is wound with very thick fibre ropes or alternatively it may be made of rubber discs threaded on to a wire rope.

To fish a trawl it has to be lowered to the seabed. It is then pulled slowly over the bottom by two warps, which are joined to the two otter boards in such a way that the pressure of the water against the boards forces them to shear outwards, thus spreading the mouth of the net horizontally. The otter boards are undoubtedly the most important

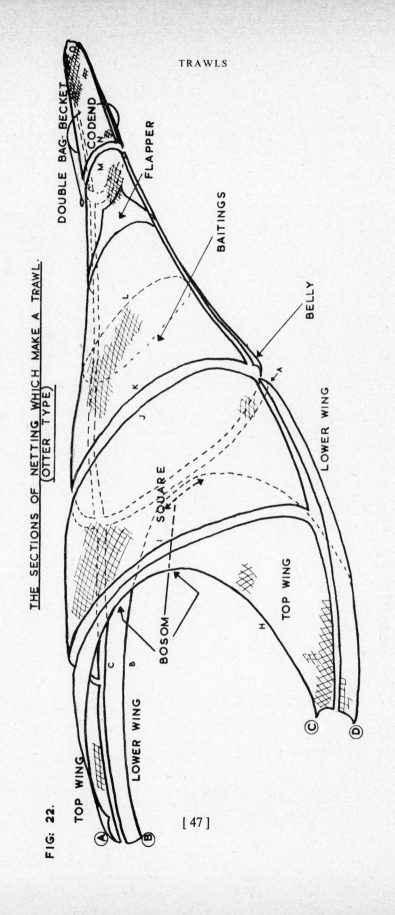

THE SECTIONS OF NETTING WHICH MAKE A TRAWL.
(OTTER TYPE.)

FIG: 22.

DOUBLE BAG. BECKET

CODEND

FLAPPER

BAITINGS

BELLY

LOWER WING

TOP WING

SQUARE

BOSOM

LOWER WING

TOP WING

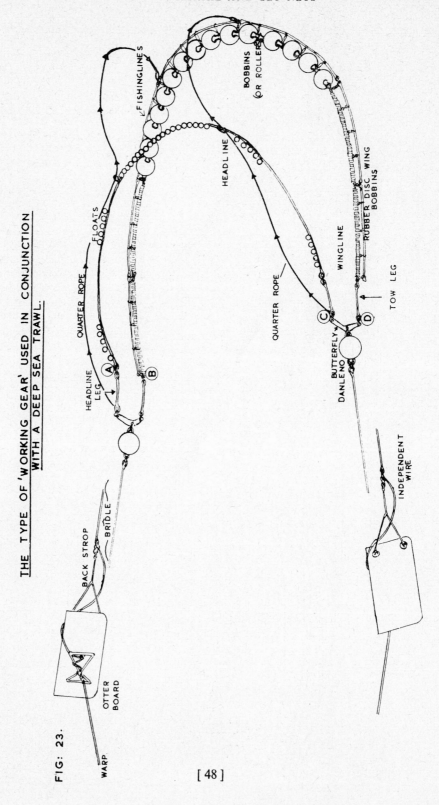

THE TYPE OF 'WORKING GEAR' USED IN CONJUNCTION
WITH A DEEP SEA TRAWL.

FIG: 23.

functional devices of any trawl, and they have to be of a correct size and weight to conform with the design, size and power of the vessel, as well as the spread and load of the type of trawl used. Many factors govern the angle at which an otter board will tow, the principal being the positioning of the brackets and the eyebolts for the backstrops. Fig. 24 gives an indication of the appropriate positioning of such fittings.

Formation of netting, attachment of floats and correct weighting cause the trawl mouth to open vertically. For converting to herring trawling secondary backstrops can be set between an eighth and a quarter.

Fig. 25 shows an otter trawl when being towed. The arrows indicate the main strain.

Improved Modification of the (Deep Sea Trawl) Granton Trawl

WITH THE possible exception of slightly different methods of adapting the working gear of a deep sea trawl, and improved handling, there has been very little change in the existing design over a number of years, especially in the net, which is principally of the Granton specification. Of the reasons for lack of improvement, one of the most important may be that any change for the better, connected with more productive fishing, is usually rather expensive. Another reason may be, whilst trawler owners are readily interested in any proposals which could lead to greater fishing capabilities, the fishermen may be inclined to regard any such suggestions suspiciously. This is probably due to the all important factor of time, for the skipper has to make the most of every available minute at the fishing grounds, which, of course, does not encourage experimentation. Again familiarity with a certain type of gear is a drawback when new ideas are introduced.

Although rather ironical, it would seem true to say that modern trawlers are designed to suit existing types of gears. The latest developments in stern trawling is a great step forward, but again the trawl has made little, if any, progress, except possibly that the trawl can be towed at a higher and more efficient speed.

One of the principal desires for an efficient trawl is to achieve as much height of headline as is possible, and one reads various claims by manufacturers of greater bouyancy for their floats. Whilst better floats are welcome, it must be appreciated that a net of specific design cannot be stretched beyond its limits. In other words, after this limit, any greater height which is achieved by improved floatation must be compensated by a proportionate 'give' in the spread, or alternatively there will be an unnatural strain on the netting.

There is little doubt that the size of the Granton trawl is ideal for present day deep sea trawling, therefore, it may not be wise to make any major changes in size, for a larger design would have some effect on the speed and probably do more harm than good. Some slight variation, however, may produce beneficial results, without adversely affecting, to any extent, the speed of tow, or necessitating any real difference in handling.

Fig. 26 shows the existing design as against the proposed design. Comparing the two it will be noticed that there is little alteration, in fact, the only slight differences are in: the square, which alters shape a little and becomes forty meshes broader at the head, in the lower wings which do not decrease in width quite so rapidly as they do in the existing specification, and the top wings where the greatest change takes place. It is here that the forty meshes increased in the square are absorbed (20 meshes in each wing) which leaves the top bosom unchanged. Also the bellies and lengtheners are made as one, to give a superior taper.

The lower bosom meshes would be ten meshes short because five meshes are transferred to each lower wing for the purpose of forming the improved corners, and this should lead to a lessening of strain, besides minimising tearing familiar to this part of the net. Similar corners are made at the top wing 'quarters'.

At this juncture it is interesting to note that the length of the headline, and for that matter the total length of the fishing lines, also footrope/bobbins, etc., remain unaltered, in

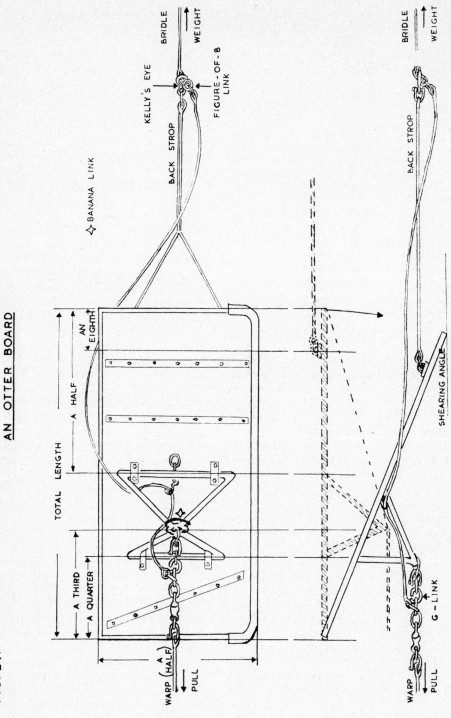

AN OTTER BOARD

FIG: 24.

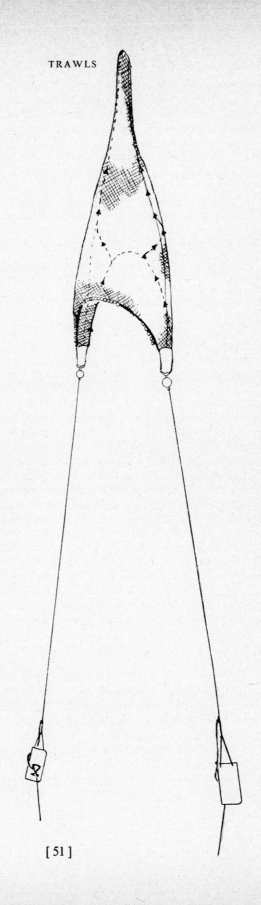

A DEEP SEA OTTER TRAWL

MAIN STRAIN ALONG FOOTROPE, VIA: BELLY TO SELVEDGES
AS SHOWN BY ARROWS

FIG: 25.

spite of the increases in the netting. The only difference would be that the last length of the wing bobbins would be unattached to the end fishing line and this in turn would support the wing end as far as the selvedge.

As for the dan leno assembly which will play an important part in increasing the efficiency of the trawl, there are two methods whereby it may be utilised. In the first instance, and this is probably the better method, the dan leno would only have two legs leading from it, as is the case with the original type of gear, only they would be considerably longer, say thirty feet, instead of ten feet. This is illustrated in the enlarged drawing. If this design was used, the quarter ropes would have to be proportionately longer i.e. twenty feet. With the second suggestion there would be three legs, an extended headline leg, an intermediate supporting leg and the normal tow leg. The intermediate leg would shackle to the end of a shortened wing line, the last eye of the supporting fishing line and finally a strengthening rope which would be attached to the original belly line and laced along the selvedge to the wing end, thus giving the headline more freedom to lift. (This is shown by the heavy line on the drawing.) With this design the quarter ropes would again have to be slightly longer than they are normally. There is no apparent reason why the introduction of this net should present any difficulties where the German 'Bang on' method is being practised.

The use of polythene netting for the altered top wings and square would undoubtedly help to accentuate the design, but this is not essential.

Theoretically it has been estimated and partially proved by models that the above variations made to the standard Granton trawl would increase the vertical mouth opening by more than seven per cent. and improve the spread of the net by a minimum of twelve per cent.

The most progressive advancement so far made in gear design, is probably that of the 'Box shaped' floating trawl, but whilst a gear based along these lines will undoubtedly play an important part in future trawling, it may be that, before it can be fully commercially successful the gear attachments would have to be more adaptable, and allow for bottom fishing in addition to mid water fishing.

Phrased differently: when floating trawls are designed in such a way that they can easily be converted into bottom trawls, say by simplified interchangeable working gear assemlies, then the floating trawl will surely replace the conventional type of otter trawl.

The Wing Trawl

Reference Fig. 27

THE WING trawl, or skagan net as it is sometimes called seems to have been introduced into Britain in 1957. This design originated in Denmark, as did the Seine net, which came into the British Isles during the 1920s. The wing trawl has far more in common with a trawl than a Seine net in that the various sections are assembled similarly to an otter trawl as will be seen by the diagram (Fig. 27) the difference being that it has more fullness, plus the shaped swallow tails between the top and lower wings, which allows the headline to rise high over the seabed.

A wing trawl can be fished in the same way as a seine net, or as a trawl with boards and steel wire. The former method seems to be by far the most popular at the present time; however, this position may alter with the trend for more powerful boats. Again wing trawls can be manufactured in sizes to suit the different types of vessels. A specification along with a diagram (Fig. 27) and a few interesting details of one of the most popular nets in use by the 114/152 horse power boats are given on the following pages.

Because it is basically a trawl, the wing trawl must be handled more carefully when shooting, than in the case of a seine net, if being fished as a seine net. This is to compensate for the more complex formation of the wings, and therefore, when paying the net over the side, the weight has to be on both pairs of wings simultaneously. It also seems advantageous to tow the wing trawl from the stern rather than the quarter.

FIG: 26. IMPROVEMENTS TO THE DEEP SEA TRAWL GRANTON TYPE

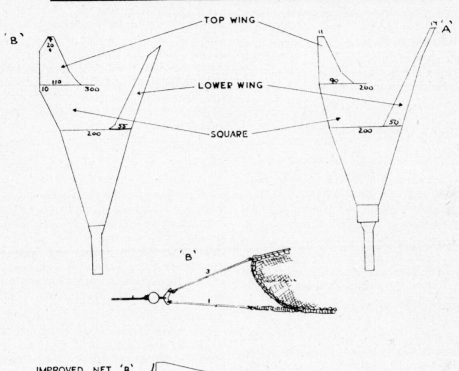

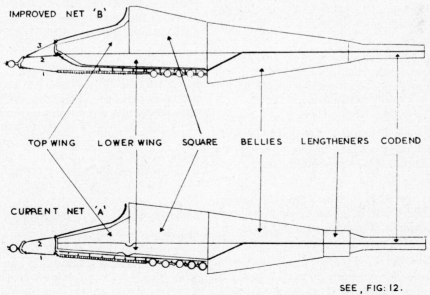

TOP WING   LOWER WING   SQUARE   BELLIES   LENGTHENERS   CODEND

SEE, FIG: 12.

It is interesting to note that since the introduction of this net, there has been a tendency, in Britain, to alter existing types of seine nets into a net which incorporates the principal assets of the trawl—whilst the formation remains basically seine—namely, the greater height, shorter and deeper wings, (and shoulders), also the inclusion of the swallow tail. This converted type of net does appear to fish better than the wing trawl, in shallow water.

Specification of the Skagen Net

Reference Fig. 27
Square

One piece of netting 226 meshes 6¼ in. mesh size bated each side every third row, down to 208 meshes in a length of 31 rows. 12s/18 ply cotton twine.

Bag No. 1

Two pieces of netting each 275 meshes 5 in. mesh size bated each side every third row down to 227 meshes in a length of 73 rows. 12s/18 ply cotton twine.

Bag No. 2

Two pieces of netting each 299 meshes 3½ in. mesh size, bated each side every third row down to 169 meshes in a length of 197 rows. 12s/18 ply cotton twine.

Bag No. 3

Two pieces of netting each 215 meshes 2¾ in. mesh size, bated each side every fourth row down to 115 meshes in a length of 199 rows. 12s/21 ply cotton twine.

Flapper

One piece of netting 162 meshes 2¾ in. mesh size, cut along a 'halfer' mesh each side down to 46 meshes. 12s/18 ply cotton twine.

Bag No. 4

Two pieces of netting each 115 meshes 2¾ in. mesh size bated each side every sixth row down to 50 meshes in a length of 193 rows. 12s/24 ply cotton twine.

Codend

Two pieces of netting 50 meshes down to 50 meshes straight in a length of 121 rows, 2¾ in. mesh size. 12s/42 ply cotton twine.

120 thread Cotton Codline meshes, bated alternately.

Chafer

For protecting Codend. One piece 60 meshes down to 60 meshes straight in a length of 38 rows 4½ in. mesh size. 75s/4 strand Manila or Synthetic trawl twine.

Set one end in six rows up from codline meshes, count 60 rows up and set in the other end, loop bated alternately round at top and bottom, setting 60 meshes of chafer into 100 meshes round codend.

Top Wings
Section 2

Two pieces of netting each 94 meshes down to 63 meshes in a length of 91 rows. 6¼ in. mesh size 12s/18 ply cotton twine. Creasings are made every 5th row on the outer selvedge. On the inner selvedge 'halfer' meshes are made, losing at 1½ meshes each time. (*See* Top Gusset Fig. 27 A and C). This has to be set into the square and top wings (As shown Fig. 27 D) and then continue along a halfer.

Section 1

Two pieces of netting each 63 meshes down to 18 meshes in a length of 47 rows. 6¼ in. mesh size 12s/18 ply cotton twine. Continue along 'halfer' on inner selvedge. Outer selvedge lose half mesh every tenth row.

Top Wing Ends

Two pieces of netting (Shown in Fig. 27 D shaded. Also shown is the two mesh guard which is braided along the entire inner selvedges of the top and lower wings.) Each 18 meshes decreasing along a 'halfer' each side down to 11 meshes. 12s/42 ply cotton twine is used for all the parts shown shaded on Fig. 27 D, i.e. GUARDS, GUSSETS, WING ENDS, ELBOWS.

Lower Wings
Section 3

Two pieces of netting each 88 meshes down to 70 meshes in a length of 35 rows, 6¼ in. mesh size 12s/18 ply cotton twine. Outer selvedge is straight. On inner selvedge nine 'halfer' meshes are made as shown in Fig. 27

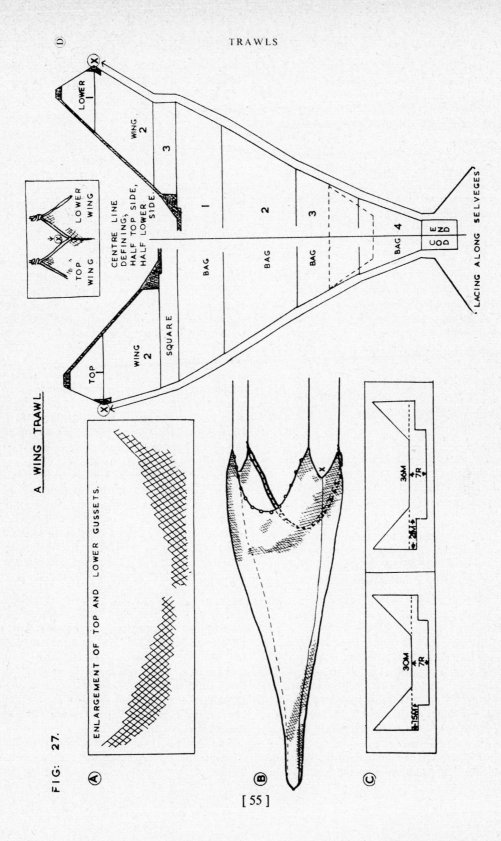

A WING TRAWL

FIG: 27.

Ⓐ ENLARGEMENT OF TOP AND LOWER GUSSETS.

Ⓑ

Ⓒ

Ⓓ

CENTRE LINE DEFINING; HALF TOP SIDE, HALF LOWER SIDE.

TOP WING 2
SQUARE
BAG
BAG
BAG
BAG 4

LOWER WING 2 3
WING 1
2
3

LOWER TOP WING WING

LACING ALONG SELVEGES

COD END

30M 7R

2M 30M 7R

b-15M/r 30M 7R

A and C (Lower Gusset is set in to bag No. 1 and the wings so that the base of each lower wing only measures 88 meshes from the outer selvedge to the beginning of the first 'halfer' mesh) and then continue along a 'halfer'.

Section 2

Two pieces of netting each 70 meshes down to 56 meshes in a length of 97 rows. 6¼ in. mesh size 12s/18 ply cotton twine. Outer selvedge one bating on the second row and five batings every third row (15 rows) repeat. Continue along the 'halfer' on the inner selvedge.

Section 1

Two pieces of netting each 56 meshes down to 15 meshes in a length of 47 rows. 6¼ in. mesh size 12s/18 ply cotton twine. Outer selvedge two batings on the second row (four rows) and one bating on the fourth row. Continue along the 'halfer' on the inner selvedge.

Lower Wing Ends

Two pieces of netting each 15 meshes down to eight meshes, decreasing along 'halfer' on each side.

Joining

The various sections are joined together as shown in Fig. 27 D. With this type of net the taper is for the most part achieved by decreasing the mesh size towards the tail.

Lacing

The complete top and lower sides are laced together, taking up two or three meshes, to position X. (As shown on Fig. 27 D.) Thus leaving the dovetail formation peculiar to this type of net.

Rigging
Headline

The netting of the top wings and square is hung into a length of 90 ft. of 1½ in. combination wire rope, with the 30 meshes of the bosom hanging loosely into 3 ft. 6 in. and the nine 'halfer' meshes into a length of 3 ft. 6 in. The remainder of the wing is marled onto the headline with the wing stretched. A 36 ft. overhang is allowed at each end.

Six 7 in. floats are secured along the bosom and quarter headline sections, and eighteen 5 in. floats are fixed along the wing sections with the spacing increasing towards the wing end.

Footrope

The netting of the lower wings and bag head is hung into a length of 110 ft. of 1½ in. combination wire rope, with the 36 meshes of the bosom hanging loosely into 3 ft. 6 in. and the 11 'halfer' meshes into a length of 4 ft. The remainder of the wing is marled onto the footrope with the wing stretched. A 36 ft. overhang is allowed at each end, dovetail between top and lower wings.

This is supported with 30 ft. of 1½ in. circumference fibre rope, divided and marled evenly between the wings. A small allowance is made at either end, which is spliced into the footrope at the lower wing end and into the headline at the top wing end.

Heaving Becket

Fifty 1¾ in. dia. plastic rings are attached around the chafer (four meshes from the top) also a thimble at each end. A 2 in. circumference steel wire rope is then passed through the first thimble, then the 50 rings and finally the other thimble. The heaving becket is then cut and eye-spliced at each end to make a finished length of approximately 12 ft.

For working the seine net method a 3 in. coir rope can be seized in bites along the 1½ in. circumference combination footrope, and weighted with about 250 seine net leads, but for trawling a weighted steel wire rope rounded with 1½ in. circumference rope or small bobbins would be more suitable.

NOTE. When joining the various sections it will be necessary in most cases to bate or— take up—the appropriate number of meshes of the smaller section. For instance, to join bag No. one to bag No. two (227 to 229) it will be necessary to take up 72 meshes (61 every third mesh and 11 every fourth mesh). In practice this can be done as follows: Make five batings every 4th then 61 every 3rd and finish with six every 4th.

The footrope is weighted similarly to the footrope of a seine net, or according to choice.

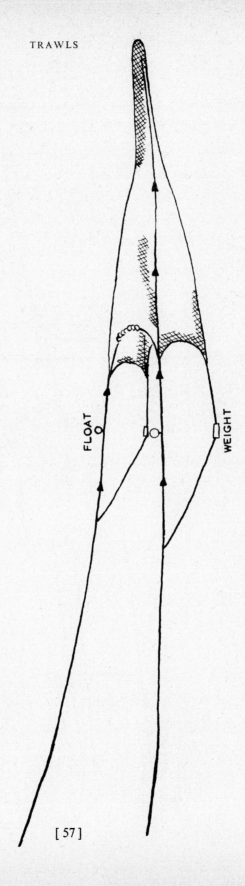

FIG: 28.

A TWO BOAT MIDWATER TRAWL

MAIN STRAIN ALONG THE UPPER SELVEDGES, AS SHOWN BY ARROWS

FLOAT

WEIGHT

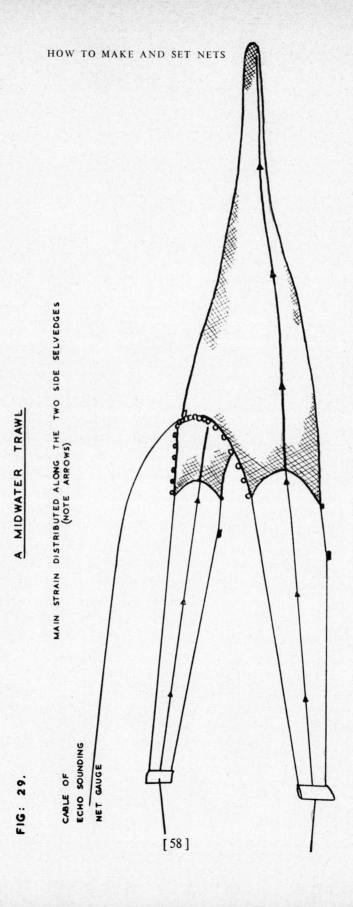

FIG: 29.

A MIDWATER TRAWL

MAIN STRAIN DISTRIBUTED ALONG THE TWO SIDE SELVEDGES
(NOTE ARROWS)

CABLE OF
ECHO SOUNDING
NET GAUGE

TRAWLS

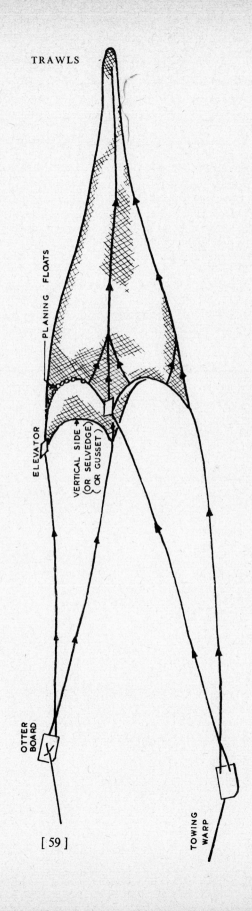

A MIDWATER TRAWL

MAIN STRAIN ALONG THE FOUR JOINING SELVEDGES AS SHOWN
BY ARROWS

FIG: 30

PLANING FLOATS

ELEVATOR

VERTICAL SIDE
(OR SELVEDGE
(OR GUSSET)

OTTER
BOARD

TOWING
WARP

The Development of Mid Water Trawls

Reference Fig. 28, 29 & 30

AT THE present time the design of trawl nets and their assembly gear is going through a revolutionary period, with the advent of surface and mid water trawls. Certain of these new designs particularly two boat mid water trawls have been proved to be commercially successful, especially when fished with smaller craft. It also seems likely that in the very near future single boat mid water trawls will be perfected to the extent where they will be in popular use commercially.

As it is now, single boat mid water trawls have, in a number of cases, proved very successful, but these successes have been largely the result of experiments carried out by certain experienced scientific groups. It does appear to be apparent, however, that the progressive steps which have been taken, and the encouraging results which have been achieved point to a future when the mid water trawl will be the net.

The major requirement for accelerating its acceptance and use in the deep sea trawling industry is the design of a net which can be fished between the surface and the sea bed, when the depth can be regulated efficiently on speed variations; at the same time it must be adapted in such a way that it can be lowered to the sea floor and fished as a bottom trawl.

Fig. 28, 29 and 30 are sketches of a few mid-water trawl designs which have already proved to be very successful. Reference should also be made to Fig. 25 with regard to strain.

CHAPTER 8

POUND NETS

Salmon trap nets, beach pound and ring nets with their detailed rigging and setting are illustrated

Salmon Trap Nets (or Pound Nets)

Reference Fig. 31, 32, 33, 35 & 36

SALMON TRAP nets, which fall in the category of pound nets, are probably among the most complicated of net designs. Looking at Fig. 31 it will be seen that a trap net consists of a bag of netting, which is formed with a top cover (N.O.P.Q.) two vertical walls (A.D.G.) which taper slightly towards the tail end (R) and a floor (M.L.K.). Vertical sheets of netting (B.C/E.F/H.I) are fitted at an angle inside the bag, in such a way that they form pockets which act as non-return valves against any fish passing through.

The mesh size decreases from 6 in. to 7 in. at the mouth, down to $3\frac{1}{2}$ in. at the tail of the net. As shown in Fig. 32 a rope is laced along every join where one section of netting meets another, for the purpose of strengthening. The framework of rope used on a large net may weigh between $1\frac{1}{4}$ and $1\frac{1}{2}$ cwts, and the usual practice is to mount the outer edges with two $2\frac{1}{2}$ in. circumference hard laid manila, while the inner ropes along the scales and hooks, which do not require so much protection, may be strengthened with $1\frac{1}{2}$ in. to $1\frac{3}{8}$ in. circumference manila rope.

The entire bag of netting is held open with wooden stretchers and flat floats secured along the frontal edge of the net, or with stakes, depending on the type of trap.

After the net has been moored or staked, a long sheet of 6 in. mesh netting, called a leader (J) is fastened near the mouth of the trap and lead towards the shore. In some instances it may taper gradually towards the shore, depending on the incline of the beach. Again the top edge is fitted with floats or staked as the case may be. The purpose of the leader is to deflect fish from their intended path and guide them into the trap.

In common with other main fishing practices, catching salmon with the use of trap nets can be a very profitable business, or on the other hand, it can be costly to the newcomer, for it is an extremely skilled and traditional mode of fishing, in which the positioning of the net is all-important. A great deal depends on the location where the nets are set, for a matter of feet can make all the difference in the catching ability. Naturally some coastlines are more open to the weather than others, and trap nets fished in such open areas are of course, more liable to damage than those worked in waters which are partially protected from the weather.

Trap nets or pound nets are used for catching many other species which swim along the coasts of different countries, and naturally they are designed in accordance with the factors conditioned by the habits of the particular species to be caught, and the location being fished. The principle, however, is generally similar, and as it would be difficult to include every design and specification in use, the three main types in use in Scotland have been selected to give the reader an idea of the working principle in addition to the main details of these nets.

Basically the assembly netting of all trap nets in use around the Scottish coastline is of a similar pattern even though there may be differences in dimensions as determined by local conditions. At the same time, however, it must be pointed out that there are three major differences in the method of setting traps and consequently these fall into three groups which are covered in the following paragraphs.

1 The Bag Net
Reference Fig. 33

This is the largest type of trap. It is fully moored with casks and anchors or rock

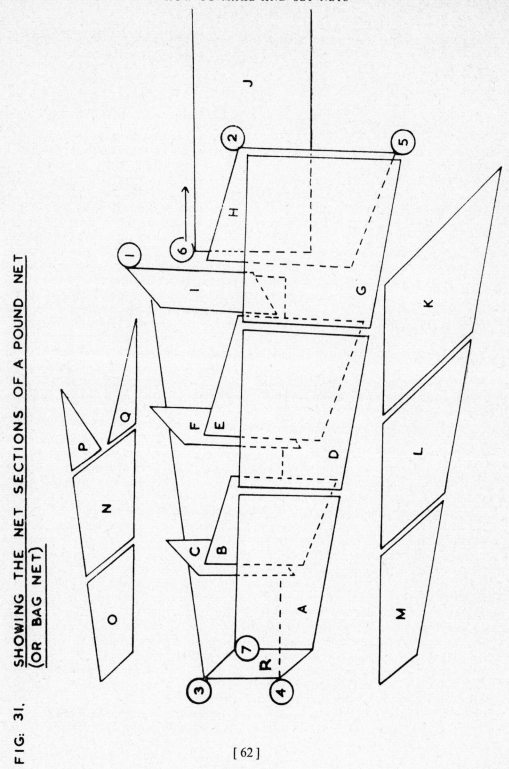

FIG: 31. SHOWING THE NET SECTIONS OF A POUND NET (OR BAG NET)

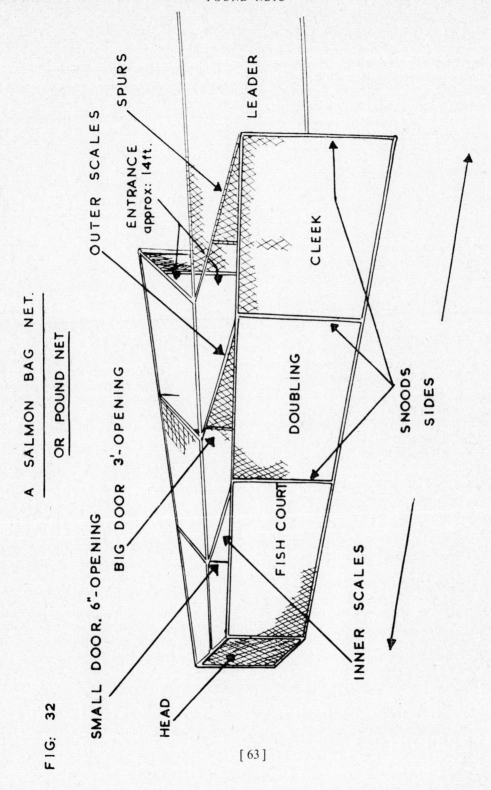

A SALMON BAG NET.
OR POUND NET

FIG: 32

SPURS

OUTER SCALES

ENTRANCE approx: 14ft.

SMALL DOOR. 6"-OPENING

BIG DOOR 3'-OPENING

LEADER

CLEEK

DOUBLING

SNOODS
SIDES

FISH COURT

INNER SCALES

HEAD

fastenings and four main poles. It is usually moored in 10 to 18 fathoms of water, so that the roof of the net and the top edge of the leader floats on the surface. Stretchers help to hold the net open, and one of these, at the rear of the trap, can easily be released when the net is fished, using a shallow cobble.

In the following the detailed dimensions of the netting specification of a standard type bag net is denoted.

Reference Fig. 31
(Also 33)

Sides
(G) 48 meshes to 39 meshes in a length of 51 meshes. 6 in. mesh of 21 thread cotton twine
(D) 58 meshes to 54 meshes in a length of 80 meshes. 4 in. mesh of 24 thread cotton twine
(A) 61 meshes to 54 meshes in a length of 98 meshes. $3\frac{1}{2}$ in. mesh of 30 thread cotton twine

Head
(R) 41 meshes by 54 meshes. $3\frac{1}{2}$ in. mesh of 30 thread cotton twine

Scales or spurs
(H & I) Outer Scales (or hooks) 46 meshes to 44 meshes in a length of 45 meshes
6 in. mesh of 21 thread cotton twine
(E or F) Mid Scales 58 meshes to 54 meshes in a length of 90 meshes. 4 in. mesh of 21 thread cotton twine
(B or C) Inner Scales 61 meshes to 50 meshes in a length of 69 meshes. $3\frac{1}{2}$ in. mesh of 21 thread cotton twine

Cover
(P & Q) Triangular Wing Pieces
(N) 100 meshes to 76 meshes in a length of 72 meshes. $4\frac{1}{2}$ in. mesh of 30 thread cotton twine
(O) 76 meshes to 32 meshes in a length of 76 meshes. $4\frac{1}{2}$ in. mesh of 30 thread cotton twine

Bottom
(K) 120 meshes to 80 meshes in a length of 44 meshes. 7 in. mesh of 21 thread cotton twine

(L) Similar to N
(M) Similar to O

Leader
(J) 46 meshes by 120 yd. 6 in. mesh

If Fig. 33 is studied in conjunction with the undernoted data, it will give an insight into the gear assembly necessary for securing purposes.

Reference Fig. 33
(C) Cleek Bridles, each leg 18 ft. long of 2 in. circumference manila
(B) 15 to 18 fathoms. $2\frac{1}{4}$ in. circumference manila
(X) 8 or 10 gallon casks
(A) 20 fathoms $1\frac{1}{2}$ in. circumference Steel Wire Rope, joining to 7 or 8 fathoms of $\frac{1}{2}$ in. circumference chain at anchor end
(Y) $2\frac{1}{2}$ to $3\frac{1}{2}$ cwt. anchor
(D) Stake
(E) Head Bridles, each leg 36 ft. long of $3\frac{1}{2}$ in. manila
(F) Length sufficient to balance head of net about one fathom
(G) 15 to 20 fathoms $3\frac{1}{2}$ in. circumference manila (can be shortened)
(H) 40 to 60 fathoms. $1\frac{3}{4}$ in. circumference steel wire rope
(I) eight fathoms $\frac{1}{2}$ in. circumference chain

An example of a standard bag net (when rigged) would be:

1 to 3	Sides ($3 \times 18'$)	54 ft.
3 to 7	Head (width)	7 to 8 ft.
3 to 4	Head (depth)	10 to 12 ft.
1 to 2	Mouth (width)	42 to 50 ft.
2 to 5	Mouth (depth)	15 to 18 ft.
6	Leader (length)	80 to 100 yd.
6	Leader (depth)	14 to 16 ft.

2 A Jumper Net
Reference Fig. 31, 34 & 35
This is about two thirds of the size of the average bag net. It is a tidal net, which is set— more often than not—with only one main stake (A) at the tail end (or head) thus leaving the bulk of the netting, including the leader, held open by seven main stretcher poles, which are in turn, held in position by ropes

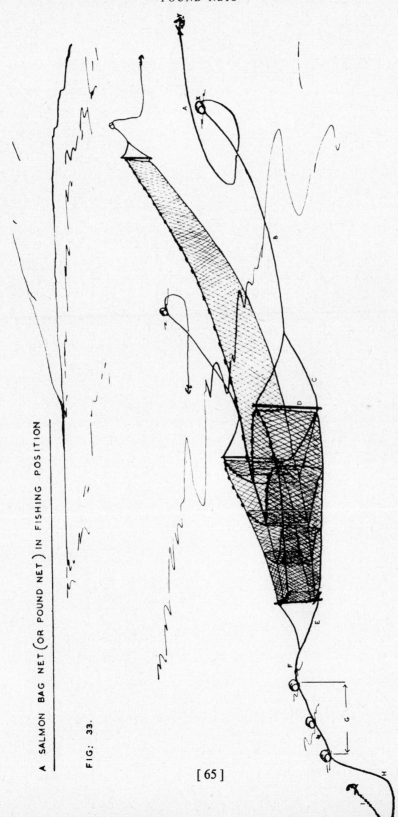

A SALMON BAG NET (OR POUND NET) IN FISHING POSITION

FIG: 33.

FIG: 34. **SHOWING A SMALL TYPE OF BEACH POUND NET CALLED A JUMPER**

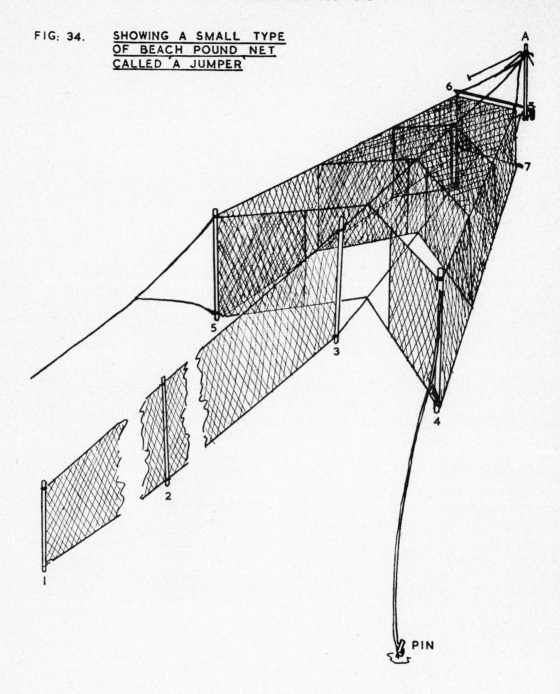

FIG: 35.

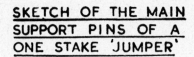

SKETCH OF THE MAIN
SUPPORT PINS OF A
ONE STAKE 'JUMPER'

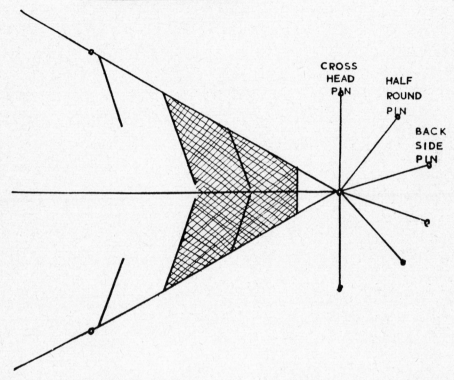

secured to pins embedded in the sand. Occasionally more than one stake is used for setting a jumper net. It is fished after the tide recedes, usually with the help of a tractor or jeep.

Referring to Fig. 31 an idea of the dimensions of a typical jumper net when hung to the ropes would be:

1 to 3	Sides (3 × 12')	36 ft.
3 to 7	Head (width)	5 to 6 ft.
3 to 4	Head (depth)	6 to 7 ft.
1 to 2	Mouth (width)	35 to 42 ft.
2 to 5	Mouth (depth)	7 to 8 ft.
6	Leader (length)	60 to 80 yd.
6	Leader (depth)	7 to 8 ft., tapering to 6 to 7 ft.

As mentioned in the notes, besides the differences in the dimensions of a bag net and a jumper net, it should be noted, when referring to Fig. 31 that the sections marked M.L.K. denote the pattern of a bag net. For a jumper net the floor of the net would be similar to the cover O.N.P.Q. This is also the case with a fly net.

3 A Fly Net

This is, in size, somewhere between a jumper net and a bag net, that is to say, the sides of the net are fastened to stakes which are firmly secured in the beach. Guy ropes leading to the embedded pins hold the stakes in position. A fly net is often set in deeper water than a jumper net, so that only part of it may be out

[67]

of the water at low tide, in which case a rope is secured around its framework in such a way that a person can walk round it to retrieve any fish which may have been caught.

NOTES. Because of the close proximity to the beach, of jumper nets and fly nets, they do not need as much floor netting as the floating bag net.

Another point of interest regarding leaders is that for most of the above nets the leaders are fitted with floats and possibly a stretcher or two, so that when the tide ebbs they lay flat on the beach, but float again into the correct fishing position, as the tide remakes. An exception to this is the fly net leader which is usually fully staked in a semi-permanent position.

CHAPTER 9

SURROUND NETS

Design and construction of the Lampara and ring nets and their methods of working

Surround Nets

SURROUND NETS are surface nets, which are designed to catch schooling fish such as herring, pilchard, tuna, sardine and salmon, etc.

The operation of encircling a shoal of surface fish with netting, varies considerably from coast to coast, and in consequence there are many surround net designs. For instance there is the two boat method when both boats of a similar size work in conjunction with a single net, or again the operation whereby a single boat independently completes the shooting and hauling of a surround net, then there is the style of fishing where a parent boat works with a skiff (or several dory's). Naturally, therefore, in order to accommodate the different ways of working, numerous net designs and rigs have been developed, so that there are in use, surround nets varying in length from 100 fathoms to 450 fathoms, or in some cases such as tuna purse seines, even as long as 700 to 800 fathoms. The depth of a net is in accordance with the power of the boat or boats, as well as the depth of the water where the net is to be fished.

The floatrope or headline of a surround net is usually made extremely buoyant, so that the upper edge of the net is prevented from sinking below the surface, a fault which would quickly release the bulk of any fish which have been caught. The groundrope on the other hand, requires only sufficient weighting to sink the net. A hawser or pursing line is attached to the groundrope with stoppers, which may be spliced or joined with metal rings, as shown in Fig. 37.

There seems to be three main types of surround nets, as undernoted:

The Lampara

The Lampara styled net, as used in South Africa, for catching pilchard. This net is of a standard mesh size throughout ($1\frac{1}{2}$ in.). It is built up with panels of netting which may number as many as nine at either side of the centre bag section. The number of meshes in the depth of each panel decreases towards the wing end by approximately 50 meshes, thus, when they are joined the difference has to be 'taken in' which effects the bagging. This is enhanced when rigging.

Purse Seines

Purse Seines, are invariably designed with large mesh wings and a centre bag of a suitable mesh size such as required for the type of fish to be caught. The exception being the net design for single boat operations, which has the large mesh at one end and the bag—where the fish will be held for brailing—at the other end. With this type of net it is often necessary to have a small lug of heavy netting, joined to the bag in order to facilitate hauling. Purse seines are, more often than not, rigged to the ropes with the wing ends squared.

Ring Nets

Ring Nets, are of a lighter structure than the average purse seine, the netting being formed with wings of a larger mesh size, shoulders of a smaller mesh size, and a centre bag. Contrary to purse seines, the wing ends of a ring net are usually gathered when rigging to the ropes, as shown in Fig. 37.

Guarding. (Selvedging)

Surround nets are strengthened around the edges with strips of heavy netting.

Rigging

There are many diversive methods of rigging surround nets, and it would be impossible to cover them all. A great deal depends upon the type of operation, for example, with the more recent developments of certain types

FIG. 36. <u>A TYPICAL RING NET</u> (75 SCORE X 330 YARDS)

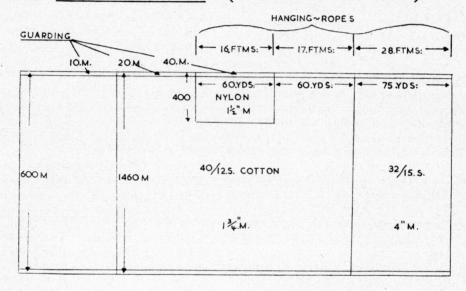

NOTE: SEE FIG. 37.

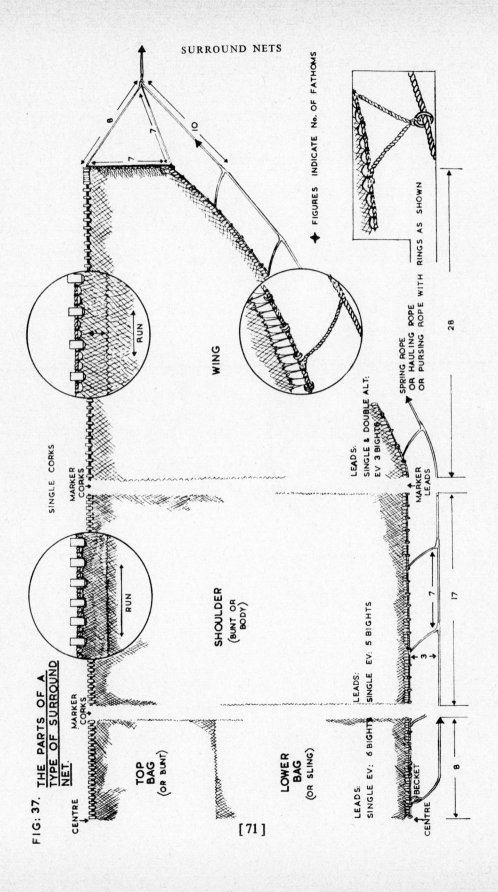

SURROUND NETS

FIG: 37. THE PARTS OF A TYPE OF SURROUND NET.

FIGURES INDICATE No. OF FATHOMS

WING

SHOULDER (BUNT OR BODY)

TOP BAG (OR BUNT)

LOWER BAG (OR SLING)

SINGLE CORKS

MARKER CORKS

CENTRE

RUN

RUN

LEADS: SINGLE & DOUBLE ALT. EV 3 BIGHTS

LEADS: SINGLE EV: 5 BIGHTS

LEADS: SINGLE EV: 6 BIGHTS

SPRING ROPE OR HAULING ROPE OR PURSING ROPE WITH RINGS AS SHOWN

MARKER LEADS

BECKET

CENTRE

FIG: 38. A METHOD OF WORKING A SURROUND NET.

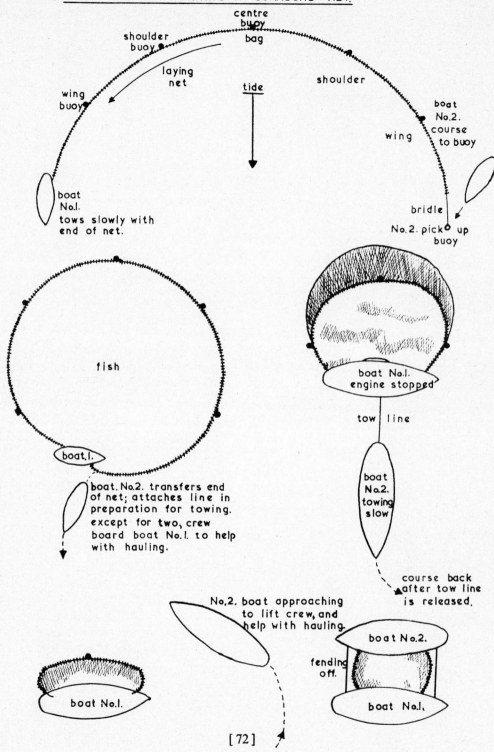

centre
buoy

shoulder
buoy

bag

laying
net

tide

shoulder

wing
buoy

boat
No.2.
course
to buoy

wing

boat
No.1.
tows slowly with
end of net.

bridle

No.2. pick up
buoy

boat No.1.
engine stopped

tow line

boat
No.2.
towing
slow

fish

boat.1.

boat. No.2. transfers end
of net; attaches line in
preparation for towing.
except for two, crew
board boat No.1. to help
with hauling.

course back
after tow line
is released.

No.2. boat approaching
to lift crew, and
help with hauling.

boat No.2.

fending
off.

boat No.1.

boat No.1.

[72]

FIG. 39. SHOWING THE BUILD UP OF THE PANELS OF A LAMPARA

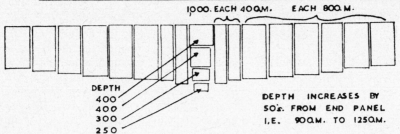

1000. EACH 400.M.   EACH 800. M.

DEPTH
400
400
300
250

DEPTH INCREASES BY
50's. FROM END PANEL
I.E. 900.M. TO 1250.M.

of power block adaptations, experiments have been made with nets which were rigged to the headline and leadline, allowing only the minimum percentage of looseness, which means that the netting is virtually stretched along the ropes, thus permitting the gear to be hauled over a block without fouling; the

To achieve speedy closure, metal rings are fitted with stoppers to the leadline. The pursing line is threaded through the rings, and when heaved inboard it draws the bottom of the net together, thus forming an artificial pond.

Not all surround nets are fitted with purs-

FIG: 40. PURSING A NET.

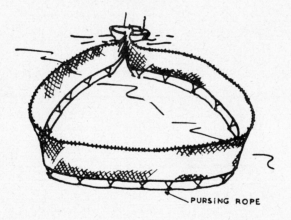

PURSING ROPE

principle being, that when the net has been laid in the water the weighting is such that the netting sinks and opens whilst the ropes adopt a snake like formation. This is, of course, one extreme method of rigging. With the average purse seine, however, one usually finds that the netting is hung looser on the headline than it is on the leadline; the difference being around 15 per cent. This of course allows the lower netting to scoop under the shoal as soon as hauling begins.

ing rings and one exception is the herring ring net which is currently in use in Britain. *See* Fig. 36 and 37. This net is fitted with the stoppers spliced into the spring rope which in effect serves a similar purpose to that of the pursing rope.

To give the reader a deeper understanding of one particular type of surround net, a ring net of the Scottish type is dealt with in the following pages.

Ring netting is a two boat operation with

50 ft. boats being a very suitable size. A popular size of net is 75 score (1,500 meshes) deep × 330 yd. stretched netting, as shown by Fig. 36. Fig. 37 illustrates how the net might be rigged.

Approximately 550 ($3\frac{3}{4}$ in. × $1\frac{1}{2}$ in. × $\frac{5}{8}$ in. bore) corks are spaced along a $1\frac{3}{4}$ in. circumference manila floatrope, whilst a 2 in. circumference manila footrope is weighted with about 400, $\frac{1}{2}$ lb. ring type leads. The spring rope may be of $2\frac{1}{2}$ in. circumference manila and the stoppers of $1\frac{1}{4}$ in. circumference manila rope. And finally, bridles may be of $2\frac{3}{4}$ in. circumference manila.

The two boat ring net operation is performed as illustrated by Fig. 38. A buoy or light (if night fishing) being positioned, after which the net is laid in a semi circular pattern by one boat, the second boat then picks up the buoy and one end of the net, before proceeding towards boat number one. When the two boats meet and the circle of netting is completed, the majority of the crew of the second boat transfer to boat number one to assist with the hauling, and a tow line is made fast—as shown. The net is hauled from both ends with the footrope gathered inboard ahead of the floatrope. This continues until the fish are gathered into a small artificial pond created by the bag of net.

By this time boat number two will have released his tow line, and will be approaching to assist in taking out the fish with brailer nets.

When hauling the main weight of the net is borne by the spring rope which is usually heaved in with a power winch.

CHAPTER 10

APPENDIX

Eighteen different types of fishing nets and gear are illustrated

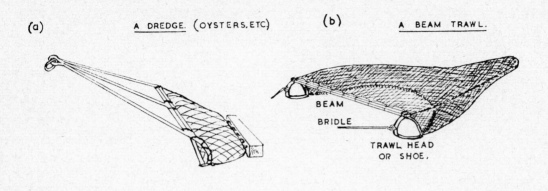

(a) A DREDGE. (OYSTERS, ETC.)

(b) A BEAM TRAWL.

BEAM

BRIDLE

TRAWL HEAD
OR SHOE.

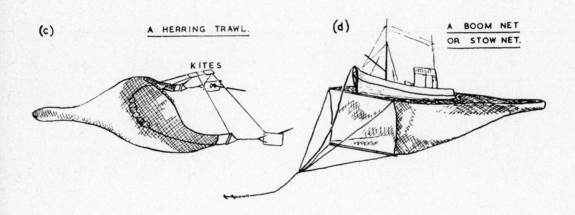

(c) A HERRING TRAWL.

KITES

(d) A BOOM NET
OR STOW NET.

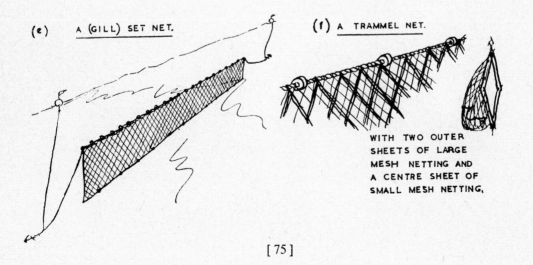

(e) A (GILL) SET NET.

(f) A TRAMMEL NET.

WITH TWO OUTER
SHEETS OF LARGE
MESH NETTING AND
A CENTRE SHEET OF
SMALL MESH NETTING.

(g)

A TYPE OF
CRAYFISH TRAP

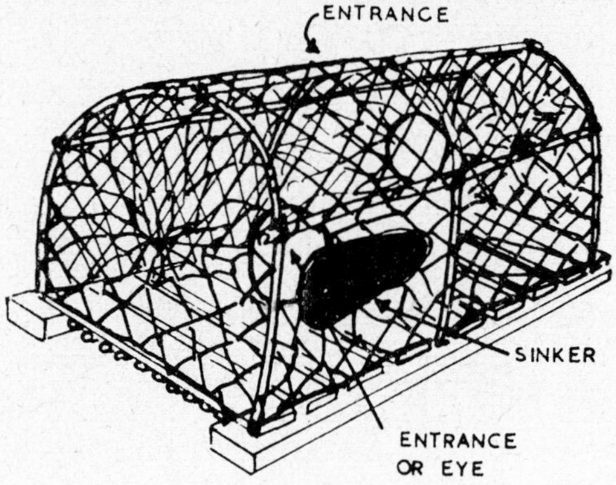

ENTRANCE

SINKER

ENTRANCE
OR EYE

(h)

A BEACH SEINE

SHORE

← TO BEACH

(i)

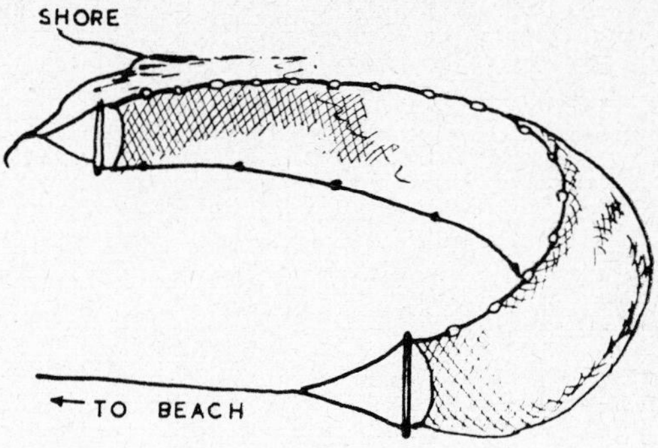

THE CORNER OF A
HERRING DRIFT NET.

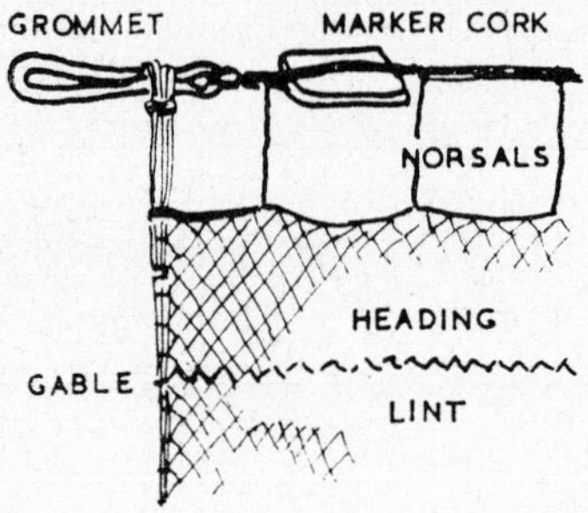

GROMMET   MARKER CORK

NORSALS

HEADING

GABLE   LINT

(j)

A TYPE OF
LIFT NET.

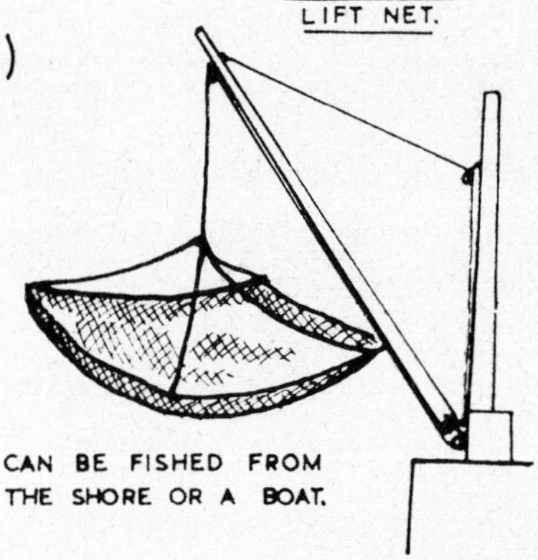

CAN BE FISHED FROM
THE SHORE OR A BOAT.

(k)

HOOP NET OR FYKE NET.

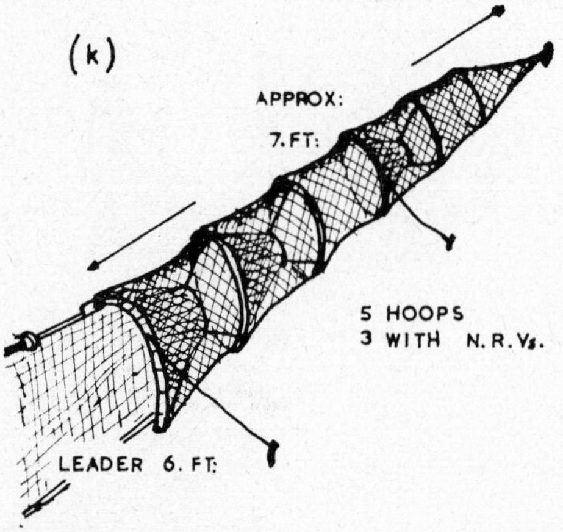

APPROX:

7. FT.

5 HOOPS
3 WITH N.R.Vs.

LEADER 6. FT.

(l)

SHOOTING DRIFT NETS.

WIND

BUOY
ROPE

BOWL OR
CANVAS
PALLETS

MESSENGER ROPE
OR WARP

STOPPER
WEIGHT

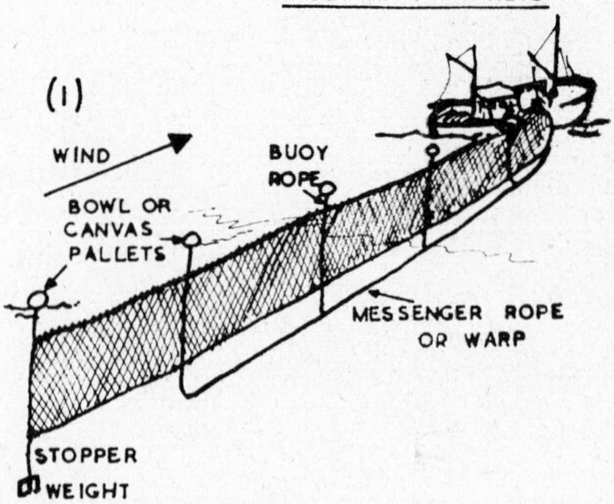

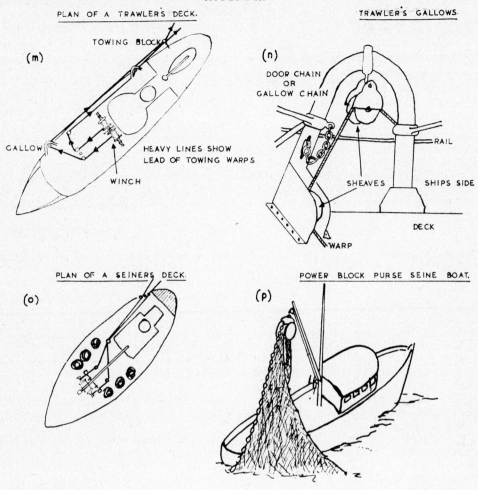

PLAN OF A TRAWLER'S DECK.

TOWING BLOCK

(m)

GALLOW

WINCH

HEAVY LINES SHOW
LEAD OF TOWING WARPS

TRAWLER'S GALLOWS.

(n)

DOOR CHAIN
OR
GALLOW CHAIN

RAIL

SHEAVES   SHIPS SIDE

DECK

WARP

PLAN OF A SEINERS DECK.

(o)

POWER BLOCK PURSE SEINE BOAT.

(p)

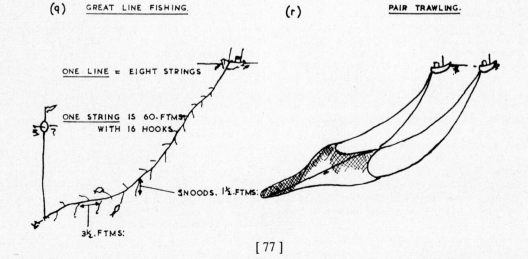

(q)   GREAT LINE FISHING.

(r)   PAIR TRAWLING.

ONE LINE = EIGHT STRINGS

ONE STRING IS 60·FTMS.
WITH 16 HOOKS.

SNOODS. 1½·FTMS.

3½·FTMS.

CHAPTER II
DICTIONARY OF FISHING GEAR AND TERMINOLOGY

Approximately 1,000 terms relating to fishing gear are listed and defined with cross references to the text

(A)

accrue. *See* creasing

air bubble curtain, type of fishing where a perforated polythene plastic pipe is laid along the sea floor to emit a curtain of bubbles, a path which the fish follow into a stop seine enclosure.

Alaska floating trap. *See* floating salmon trap

apron, the drape on the lower lip of a salmon bag net. Also called: stomach piece. *Reference Fig. 31-κ*

armouring. *See* wall

atom trawl. *See* floating trawl

(B)

back strop, a wire rope which fits to the otter board, is partially responsible for determining the angle at which the board will tow. *Reference Fig. 23*

bag, (1) the centre part of a seine net between the shoulders and the cod end, or inclusive of the cod end, or only the cod end; **(2)** when the cod end is heaved out of the water to form a bag of fish; **(3)** in ring netting or pursing the section of netting where the fish will eventually be held for brailing. Also called: bunt, sack, sling. *Reference Fig. 37*

bag net. *See* pound net

bag rope, when trawling a preventive rope fitted above the rail, for limiting the distance the cod end can swing inboard

baiting. *See* bating; also baiting piece

baiting piece, the section of a trawl between the square and the upper section of the cod end. Also called: baiting, bating, bating piece. *Reference Fig. 22*

balch line, a soft intermediate rope of low breaking strain, fitted between the netting and main working ropes. Also called: bolch line, hanging line. *Reference Fig. 18-Ex. 8*

balloon net, high opening shrimp trawl, made of lightweight webbing

banana link, a metal link fitted to the brackets on an otter board. It prevents 'G' link assembly being fouled. *Reference Fig. 24*

bang up gear, arrangement of the lower edge of a trawl, (from the toe of the dan leno to the bobbins) whereby it can be heaved inboard simultaneously

bar, one of the four sides of a mesh. Also called: half mesh, leg. *Reference Fig. 1*

bark. *See* barking

barking, process of treating nets, ropes and other materials produced of natural fibres with a protective mixture composed in the main, of tanning bark obtained from the bark of various species of trees. Also called: cutching, tannage, tanning

barking pot, iron pot in which materials are treated with bark

barking vat, wooden vessel in which materials are treated with bark

barrel. *See* gipsy

barricade, a barrier usually of non-textile material which is erected to prevent the escapement of fish after they have entered the tail of a river etc., say with the rising tide. Also called: brush weir, corral fish garth, shore weir, trap, weir

basket, for long-lining, a wicker basket may be the receptacle for the lines. It has to be fitted with a cork or soft rope rim for holding the hooks. One complete long-line consists of a number of baskets. There are also many other uses for the various range of sizes in which wicker baskets are made. The receptacle may also be called: a line, a piece, tub, unit. *See* skate

bass rope. *See* ground rope

bating, (1) in net making the process of reducing the number of meshes by taking into one mesh—two meshes of the preceding row. Also called: doubling, narrowing, shrinking, stealing, stole, stolers, stowing, taking in/up; **(2)** *See* baiting piece

bating piece. *See* baiting piece

bats, triangular pieces of netting fitted between the body and the wings of a shrimp trawl. Also called: corners, jibs

beach seine, a seine type net, often with the netting hung loosely to cause bagging, it is operated from the shore or a river bank, in a circular action. Also called: drag seine, ground seine, haul seine, shore seine, sweep net, yard seine. *See* Chapter 5- Hanging surround nets; also Appendix (h).

bead bobbin. *See* roller

beam trawl, a conical shaped net held open by a horizontal beam which in turn is fitted at each end with an iron framework which holds the beam above the seabed, thus forming the mouth of the net. Also called: drag net, pole net, pole trawl, trawl beam. *See* Appendix (b)

beating. *See* braiding

becket, (1) a length of manila line by which each gauging is fastened to the main line of a halibut long-line; **(2)** when the two ends of a length of fibre or wire rope are spliced together forming an endless becket, or when a loop splice is made and an end is left for tying—called a becket with a tail

becket bobbin, a small cylindrical iron or wooden bobbin with becket attached for securing ground rope or rollers to the lower lip of a trawl. Also acts as spacing between main rollers

bellies. *See* belly

belly, the lower central netting section of a trawl, or the entire centre portion (belly and baiting). Also called: bellies. *Reference Fig. 22*

belly line, a rope extending from the quarter junction across the bars of the belly meshes, and along the outer edge (lastridge) to the end of the cod end. It carries the strain on these sections. Alternatively a rope extending from a point on the lastridge to the end of the cod end, and connecting to the quarter junction with a wire or chain line leg. Also called: rib line, ripe line. *Reference Fig. 12*

belly line leg. *See* belly line

bight, loop of a rope. Also called: bite

bite. *See* bight

blanket net, a lift net, operated with the aid of lighting or bait. The float line is attached to, or suspended from, an outrigger pole of the boat. It is hauled by means of lines attached to the underside of the net

blinder. *See* chafer

board, *See* otter board

boarding the net. *See* hauling in

bobbin. *See* roller

body of the net, the centre which is usually the main part of a net or sections of a trawl, for instance, the bag portion of a seine net between the wings and the cod end, consisting of belly and batings. *See* shoulder 2. *Reference Fig. 13*

bog, alloy float

boggey. *See* molgogger

bog line, rope on which bogs are threaded

Bohol trap net, net into which fish of the reef species are frightened by a scareline. Also called: kayakas

bolsh line. *See* balch line

Bonito style fishing, fishing with live bait carried in tanks of circulating sea water, fitted on the stern deck of the vessel. Also called: jack pole, Jap pole, live bait, striker method, squid method

boom net, a shrimp net held open by an arrangement of wooden spars and shaped to fit around the hull of the boat using the net. The vessel anchors and the net is fished with the tide. *See* Appendix (d)

bosom, the centre part of the curve formed by the headline or ground rope of a trawl or seine. Also called: in trawls: busom, in seines: busom or crown. *Reference Fig. 22 & 19*

bosom meshes, the clean meshes between the wings of a trawl or seine net

bottom line, (1) *See* set line; **(2)** *See* ground rope

boulter. *See* throat line

bow poles, in trolling, a pair of secondary main outrigger troll poles, situated forward of the mast

bowel. *See* meshstick

bowl, (1) one of a set of small buoys attached at intervals to the balch or headline of a drift net, or fleets of drift nets. Also called: buff, pallet, pellet. *See* Appendix (1); **(2)** *See* fish court

box line. *See* long-line 1

box net, (1) trap net set through holes under ice of frozen lakes and estuaries; **(2)** *See* box type otter trawl; **(3)** a rectangular framework of netting, with three vertical sides moored with stakes and anchors. It is fitted with a leader, which guides fish into the open side. This can be raised after a shoal has entered the enclosed area

box-type otter trawl, a trawl, shaped like a box with wings attached to the sides and with a sack or tunnel fitted. Also called: box net, box-type trawl, four-piece trawl, Pacific trawl, Western trawl

box-type trawl. *See* box-type otter trawl

bracket, two triangular brackets are fastened to each otter board. The towing rope connects to these and in conjunction with the back strop etc., when towed, it diverts the board at an angle for spreading the trawl. *Reference Fig. 24*

braiding, making netting by hand. Also called: beating, weaving

brail, (1) short stick fitted at each end of a beach seine for spreading purposes; **(2)** a dip net for transferring fish. *See* brail net

brail line, (1) *See* tag line; **(2)** *See* breast purse line

brailer. *See* brail net

brailing, (1) in purse seining, the action of bringing the lead line up towards the water surface; **(2)** the operation of transferring fish surrounded by netting from the sea inboard, by means of a smaller net

brail net, small dip net for scooping out portions of the catch from the main net and transferring inboard. Also called: brailer, hand brailer

brail rings. *See* breast rings

branch line. *See* gangen

breast line. *See* breast purse line

breast of net, each end of a purse seine

breast purse line, a light line from the lower corner of a purse seine, passing through the breast rings to the upper corner, used for hauling the lead line to the surface. Also called: brail line, up and down line

breast rings, rings which are secured along the breast line of a purse seine and through which the breast purse line is threaded. Also called: brail rings

breast strip, a vertical piece of heavy netting at the ends of a purse seine

bridle, (1) one or more ropes of wire or fibre attached between the end of a net, seine or trawl, and the main towing ropes or otter boards. They act to some extent in directing fish into the tract of the net. Also called: ground cable, ground warp, sweep, sweep line. *Reference Fig. 23*; **(2)** length of chain or wire or fibre rope with both ends secured, for example the ropes connecting the pursing ring to the foot of a purse seine net

brush weir. *See* barricade

buff. *See* bowl 1

buffow line. *See* throat line

bull. *See* stapling

bully net, a conical bag of netting fitted to a circular hoop which is secured to a pole. The hoop is dropped over a crayfish and the netting is closed by means of a cord

bultow. *See* throat line

bunt, (1) in a trawl it is the section of netting at the wide end of the lower wing. *Reference Fig. 12*; **(2)** when a net is gathered to form a pocket for easier removal of fish. *See* bag 3

bunt bobbins, the bobbins along the bunt. *Reference Fig. 12 & 23*

buoy trap, a moored trap with two pairs of non-return valves, similar to a pound net but with the leader at right angles to the trap

busom. *See* bosom

bush rope. *See* herring warp

butterfly, iron spreader attached to dan leno bobbin. *Reference Fig. 23*

(C)

cable. *See* bridle

cage roller. *See* molgogger

capelin seine, a seine net used in areas north of Canada

cask, (1) *See* keg; **(2)** a small wooden barrel float used for mooring trap nets

hast, (1) throw of net, fishing line, etc.; **(2)** the fly, cook and gut

cast net, conical net usually operated by one man, which is thrown to cover the fish. Heavily weighted around the perimeter it is provided with draw cords and a retrieving line which passes through the apical portion. Also called: throw net, trow net

chafer, replaceable material, such as used netting, which is attached to the underside of the cod end to protect the main netting from wear as it slides along the sea bottom, or chafes against the side of the vessel. Where extensive abrasion is expected very hard wearing materials like cow hides may be used as additional protection. Also called: chafing gear, chafing piece, cod end protector, hula skirt

chafing gear. *See* chafer

chafing piece. *See* chafer

chain leadline, chain used on purse seines instead of ground rope. Also called: foot chain

chain line. *See* ground rope

chinchola seine, type of beach seine

chum. ground or alive fish thrown over board to draw fish. Known as chumming. This type of bait is also called: stosh, live-bait

clam dredge. *See* dredge

clam rake. *See* dredge

clean join, where two sheets of netting are connected with clean meshes

cleek, (1) the leading vertical edge (mouth) of a pound net. *Reference Fig. 32*
(2) one of the three sections which make up one side of a pound net. *Reference Fig. 32*

clip link, a metal ring with slot for connecting similar links

cod end, end of a trawl or seine net which acts as the receptacle for fish caught in the net. It is closed and held, at the extreme end, with a line. *See* cod line. Also called: bag, fish bag, pocket, tail. *Reference Fig. 12 & 22*

cod end protector. *See* chafer

cod end rope. *See* cod line

cod line, line at the extreme end of the cod end of a trawl or seine net. It is reeved through the meshes which are closed together. Fastened with a special type of slip knot which releases easily to allow unloading of fish inboard. Also called: cod end rope, draw line, draw rope, 'G' string *Reference Fig. 12*

combination rope, a mixed rope, of vegetable and fibre strands

copper treatment of net, the treatment of nets with a chemical substance containing copper as a preservative—such as copper naphthenate

coquina scoop, box with a heavy screen bottom used to catch the donax or coquina

coracle net. *See* trammel net

cork buoy. *See* dipsey

cork line. *See* headline

cork rope. *See* headline

cork strip, horizontal reinforcement of heavier netting along the upper edge of a net

corners. *See* bats

Cornish pilchard or sardine seine, small mesh encircling seine net which is laid across a bay to block retreat of a fish school

corral. *See* barricade

cover net. *See* cast net

cow hide. *See* chafer

crab dredge, triangular iron frame with a meshed bag consisting of steel rings on the lower side and fibre netting above. A drag bar with iron teeth is attached along the front lower edge. *See* dredge

crab hook. *See* crayfish hook

crab pot. *See* crayfish trap

crab scrape. *See* dredge

crab trap. *See* crayfish trap

craft lift net, framed shallow lift net which is baited, sunk to the bottom by lines and weights, and occasionally hauled quickly to the surface

crate. *See* live box

crawfish hook. *See* crayfish hook

crawfish net, (1) *See* bully net; **(2)** *See* crayfish trap

crawfish pot. *See* crayfish trap

crayfish hook, a large hook attached to a pole for hooking crayfish from crevices. Also called: crawfish hook

crayfish trap, there are many varied types consisting of a weighted framework and covering, all made to such a size that they may be fished by hand. They are anchored at, or near, the seabed. *See* appendix (g)

(1) Creel, box shaped or rectangular, with rounded top. The frame may be covered completely with heavy netting which has a tunnel leading in from either side, or covered with horizontally slatted wood with tunnel of netting leading in from the ends

(2) Pot, may be hemispherical or barrel shaped. The former type is usually of a wicker work structure, with a tunnel through the top, whilst the other may be of horizontal slats or wire mesh. The nouns: crab, crawfish, crayfish, lobster, etc., precede the descriptive phrases, and vary somewhat in different areas, i.e. crab pot, lobster creel, lobster pot and so on

creasing, when the number of meshes in hand braiding is increased one, by making an extra loop in the preceding row. Also called: accrues, false meshes, gaining meshes

creel. *See* crayfish trap

crib. *See* fish court

cross tree. in trollers, cross pieces near the top of the mast, notched to saddle the main outrigger troll pole when not in use

crown. *See* bosom

cutback, cut or hang of a trawl so that the ground rope rides behind the headline. Also called: overhang, setback, undercut, sharks mouth *See* Chapter 6—Rigging

cutching. *See* barking

cut meshes, meshes which are with the 'run' or 'scone' of the net. *Reference Fig. 2*

cut splice, two rope ends spliced so as to form a slit

(D)

daffins. *See* norsals

dahn, type of radio buoy for marking the location of fishing gear

dan, small buoy made of wooden or cork squares strapped together, supporting a pole with a flag

dandy. *See* dandy winch

dandy bridle, in beam trawling a rope fastened to the end of the beam for hauling

dandy winch, a small steam capstan or hand winch used by sailing trawlers for hauling. Also called: dandy wink

Danish Seine. *See* seine net

Dan Leno, (1) wooden spreader used at the leading end of the wing of a trawl or seine, to keep the net spread vertically. Also called: dhanoleno, spreader, the stick, trail, stretcher. *Reference Fig. 19*

(2) steel bobbin with attachments, butterfly, links, shackles, spindle, swivel, for spreading and lifting trawl mouth. *Reference Fig. 23*

dan leno board, small otter board used in place of a dan leno. Also called : pony board

dan line, rope used by fishermen for fastening a dan buoy to the lines or to a small anchor. Also called: dan tow, dan wire, keg line

dan tow. *See* dan line

dan wire. *See* dan line

davit loop, the loop at the bag end of a purse seine

deep water trap. *See* submarine trap

deeping. *See* drift net

deep-sea seine. *See* seine net

deep-water clam dredge. *See* dredge

demersal seine. *See* seine net

depressor, a shearing plane, fitted to a type of mid-water trawl, for depressing the lower edge of the mouth

depth, applied to netting, it means the number of cut meshes. *See* net depth. *Reference Fig. 2* Chapter 2

dhanoleno. *See* dan leno

diamond. *See* point

diamond hanging mesh, when each mesh is hung to a rope by one corner only, to form a diamond shape. *Reference Fig. 17*

diddle-net. *See* doddle net

digger, an instrument for digging clams. Also called: rake

dip net, a small mesh bag, sometimes attached to a handle, shaped and framed in various ways. It is operated by hand or partially by mechanical power, to capture the fish by a scooping motion. Also called: scoop, scoop net

dipsey, the float of a fishing line. Also called: cork buoy

disk roller. *See* roller

doddle net, a dip net used in the North Sea to take fish out of a beam trawl when an unusually large catch has been made. Also called: diddle net

dogears, triangular pieces of netting fitted into the angle formed by the forward edge of a balloon trawl, having all bars along the hanging end and points on the wings

dog line. *See* whiskey line

dolphin striker. *See* fish jig

door, (1) *See* otter board; (2) the opening or apex between scales in salmon trap nets, i.e. between the outer scales is the big door and between the inner scales is the small door. *Reference Fig. 32*

doors end, the leading end of the trawl-net wing

dormant meshes, the meshes between the norsals, in a wall-net. *Reference Fig. 17*

dory hand-lining, hand lining by two men in small boats called dories, which are carried by a parent boat

double bag becket, a becket secured at a suitable position around the cod end of a trawl. It is used for lifting the catch of fish inboard. Also called: halving becket, lazy rope, splitting strop. *Reference Fig. 22*

double dragging, two boat fishing with a paranzella net

double selvedge, where the edge meshes of a section of netting are formed with double twine to give extra strength. Also called: straight selvedge

doubling, (1) *See* bating; (2) the middle section of the three which comprise the side of a pound net. *Reference Fig. 32*

drag, to pull the gear through the water. Can also indicate that one operation of shooting, towing and hauling has been completed, i.e. A drag. *See* haul

dragging. *See* trawling

drag-line. *See* tag line

drag-net, (1) net in which the capture of fish is effected by a horizontal pulling or dragging motion of gear. Also called: pull net; (2) *See* beach seine; (3) *See* beam trawl

drag rope. *See* towing rope

drag seine. *See* beach seine

draught of fish, a haul of fish

draw line. *See* cod line

drawrope. *See* cod line

dredge, fishing gear consisting of a rectangular frame supporting a bag of fabric, wire webbing etc., which is dragged along by a raking or scratching action at the bottom of rivers, lakes and seas, to capture fish. Also called: clam dredge, clam rake, rake, scraper, shank net. *See* appendix (a)

dried up, when a catch of fish is ready for brailing

drift line, (1) a long line attached to a boat or a series of buoys which is allowed to drift with the natural elements; (2) a fishing line used for the capture of pollack and conger. It is a single-hook line without sinker which is allowed to drift from an anchored boat; (3) *See* long line

drift net, a rectangular gill net. Several of these nets joined end to end, form a 'fleet', which is fixed to a vessel. The vessel and nets then drift with the natural elements. Also called: deeping. *See* appendix (1)

drive in net, closing pouch-net into which fish are driven by scare devices, the mouth is then closed to prevent escape

driving net. *See* whammel

drop line, (1) a long line set diagonally between the surface and the bottom of the sea. Also called: single anchor line, up and down line; (2) *See* single hand line

drop net, net which is dropped or allowed to fall without a preceding casting movement

drum end. *See* gipsy

dry hand-lining, hand lining from deck of a vessel

dry weir, a weir which is high and dry at low tide

dual fin otter board, type used in one specific kind of mid-water trawling

dummy, a small cylindrical iron or wooden bobbin which acts as a spacing between rollers of a ground rope

(E)

encircling gill net, gill net payed out in a circle or an arc of a circle, and the gilling process is hastened by frightening the fish with various devices

endless becket. *See* becket

entangling net, net which effects the capture of fish by entanglement. *See* trammel net

eye. (1) part of hook to which line is attached; **(2)** the loop formed when the end of a rope is spliced into itself; **(3)** *See* non-return valve

eye splice, when the end of a rope is turned to form a loop then spliced into itself

extension piece, a section of netting used for increasing the length of a trawl net bag. Also called: lengthener. *Reference Fig. 26*

(F)

fair lead, a device for guiding the trawl towing rope as it pays out, to or from the winch

fair meshes, along the clean meshes. *Reference Fig. 2*

false bellies. *See* chafer

false meshes. *See* creasing

fathom, a nautical measure equal to 6 ft.

fence. *See* leader

Figure of eight link, link attached to independent piece and bridle. The purpose of this link is to facilitate connecting and disconnecting otter board, by jamming in the kelly's eye. Also called: stop link. *Reference Fig. 24*

filter net, a bag net without non return valves, set in flowing water to effect capture by straining

fine line, a lighter and shorter line than the normal long-line, with small hooks closely spaced on shorter and lighter gangens

finishing three legger. *See* halfer

fish bag, (1) *See* cod end; **(2)** *See* bag

fish court, the walls of netting forming the holding chamber in a trap net. Also called: bowl, crib, pocket, pot, pound. *Reference Fig. 32*

fish garth, a dam or weir in a river or on the sea shore for keeping fish or taking them. *See* barricade

fishing gear. *See* gear

fishing line. (1) *See* lining; **(2)** Combination rope fitted to a trawl between the balch and the ground rope or bobbins

fish jig, a hand line attached to which is a metal rod or weighted line, with several hooks set back to back. Also called: dolphin striker, fish-line jigger

fish line jigger. *See* fish jig

fish net. *See* net

fish tackle, a three fold purchase which is hooked into the double bag becket for heaving a loaded trawl net cod end inboard

fish trap. *See* pound net

fish weir, (1) *See* fish garth; **(2)** *See* barricade

fish wheel, mechanical device used for catching salmon

flaking. *See* fleeting 2

flapper. *See* non-return valve

fleet, the total number of drift nets shot by one vessel

fleeting, (1) the procedure in which a number of fishing vessels are operated together as one organized fleet; **(2)** with netting, it means folding the netting evenly in a continual operation over the previous fold. Also called: flaking

fleet rope. *See* herring warp

flicker, a tuft of feathers at the end of a light bamboo pole to excite a school of albacore

float, (1) any buoyant object attached to nets of fishing lines to prevent sinking; **(2)** *See* live box

floating line, a long line made buoyant with floats so that it lays on the surface. May be used for catching salmon

floating salmon trap, a trap used for catching salmon. It consists of nets which are suspended from a floating framework of logs fastened together. It is weighted with rocks. Also called: Alaska floating trap

floating trawl, a net which is towed between the sea bed and the surface. May be conical shaped with four small wings, or made with wings and vertical walls of netting which taper towards the tail. The depth at which the net fishes is determined by length of towing rope and speed of vessel. Also called: atom trawl, Larsen trawl, mid-water trawl, pelagic trawl, phantom trawl. *Reference Fig. 28, 29 & 30*

float line, (1) trolling line carried from the middle part of the pole; **(2)** *See* headline

flue. *See* headline

flymesh, when hand braiding netting, especially wing sections, it is the method whereby the last mesh of the previous round is omitted, thus making a series of flymeshes or hanging meshes, down the edge. *Reference Fig. 18*

fly net, (1) a drift net with no strengthening selvedge; **(2)** *See* pound net

foot chain. *See* chain leadline

foot line. *See* ground rope

foot rope. *See* ground rope

footrope roller. *See* roller

fore bay. *See* heart

four piece trawl. *See* box-type otter trawl

frame trawl, a trawl net of conical shape where the mouth is kept open by a rigid frame

funnel, (1) *See* throat; **(2)** the funnel shaped opening to a fish trap; **(3)** in trawling the strip of netting holding the double bag becket in position. Also called tunnel

fyke net, (1) a funnel-like bag net similar to a hoop net but with wings; **(2)** tubular bag, for catching eels. Held open with hoops fitted with non return valve and leader. Also called: hoop net. *See* appendix (k)

(G)

gable, the side cord attached to a herring drift net. Also called: gable end. *See* appendix (i)

gaff pole. *See* jigger pole

gaining mesh. *See* creasing

gallow, horse shoe framework, fitted with sheaves, mounted over the gunwale to facilitate working a trawl, *See* appendix (m) and (n)

gallows block, block mounted on the gallows to facilitate handling the trawl. *See* appendix (n)

gallows chain, chain fixed to the gallows used to hold the otter board in place when it is hauled up. *See* appendix (n)

gallows sheave, the sheave in the gallows block. *See* appendix (n)

gangen, a branch line by which a hook or bait are attached to the main line of a long line. Also called: branch line, ganger, gangion, ganging, gauging, lanyard, leader, snood, snell, snoot line, stageon, tom, vertical line. *See* appendix (q)

ganger, *See* gangen

ganging, *See* gangen

gangion, *See* gangen

gauge, *See* meshstick

gauging, *See* gangen

gear, (1) the tools, implements or appliances used in fishing, exclusive of the boat or boats employed in the fishing operations; **(2)** a net and its activating devices when in the water

gill net, a net (usually rectangular) with the size of mesh such, that when fish strike the net they become jammed. The mesh size naturally varies for different species, very small for sprats and herring, large for skate and shark, etc. *See* appendix (e)

gipsy, a small auxiliary drum with flanges at each end, fitted outside the big frame of a winch or windlass for heaving on running gear, hawsers, etc. Also called: barrel, drum end, gypsy head, gypsy, nigger head, shipping drum, warping end, whipping drum, winch end, winch head

gipsy head. *See* gipsy

'G' link assembly, an arrangement of metallic links used for attaching towing ropes to otter boards. *Reference Fig. 24*

grab, an arrangement capable of being cocked open which is used to catch crayfish. It is fastened to the end of a pole

grain, a type of spear used for the capture of crayfish

Granton trawl, an otter trawl. Type used for deep sea trawling. *Reference* Chapters 3 and 9

grapnel, three or four large metal hooks set together back to back; **(1)** may be used as a small anchor; **(2)** with a line attached it is used for picking up or grappling for ropes or other items, i.e. a dan line

grass rope, a coir rope. *See* ground rope

great line, a type of long line up to 14 miles long, fitted with several thousand large hooks, which are attached along widely spaced gangen. Used for cod, conger, skate and halibut. *See* apendix (q)

great lining, (1) fishing with great lines. *See* great line; **(2)** hand lining in depths of over 60 fathoms

grommet, a hoop made with wire rope. It has many uses, amongst them, when served with trawl twine, two grommets are used to hold the double bag becket in position. *See* appendix (i)

ground cable. *See* bridle 1

ground line, (1) *See* main line; **(2)** *See* gound rope; **(3)** *See* set line

ground rope, (1) a heavy wire combination, manila or coir rope, attached to the lower lip of the mouth of a trawl net. Also called: bass rope, bottom line, chain line, foot line, foot rope, grass rope, ground line, lead line, sole rope; **(2)** the lower weighted rope of a beach seine, set net, purse seine net, etc. Also called: lead line etc. *Reference Fig. 19 & 23*

ground seine. *See* beach seine

ground warp. *See* bridle

'G' string. *See* cod line

guarding, (1) *See* selvedge; **(2)** *See* heading; **(3)** *See* reining

gurdy, hand or powered spools or reels, used in trolling, they are controlled individually so that any one main line may be payed out, reeled in, or held. Also called: hurdy gurdy, line hauler, line roller

gusset, (1) in a seine net it is a strengthening patch of netting fitted between the shoulders. Sometimes fitted to act as cutback. *Reference Fig. 20 & 21*; **(2)** tapering strip of netting fitted between the top and lower sections of a floating trawl. *Reference Fig. 30*

Guthrie cod end, special type of cod end with soft multiple cords between adjacent knots to entangle shrimp and prevent their escape

Guthrie net, *See* Guthrie cod end. A trawl made using a similar principle

gypsy. *See* gipsy
gypsy head *See* gipsy

(H)

haddock line, a small set line used for catching haddock and whiting
halfer, when two bars of the meshes in a net run in a consecutive line. A halfer is necessary for starting the repair of a net, also for finishing. Also called: a three legger
half mesh. *See* bar
half ring, a term employed when purse rings are placed only half way round the headline
half ring net. *See* lampara
half round lobster pot. *See* crayfish trap
halving becket. *See* double bag becket
hand brailer. *See* brail net
hanging, (1) joining sections of webbing to form a net; (2) connecting webbing to ropes. Also called: mounting, rigging. *See* Chapter 5
hanging line. *See* balch line
hanging mesh. *See* flymesh
harpoon, spear like instrument with detachable barbed head from a shank to which a line is attached; it is thrown by hand or fired from a gun
haul, (1) the fish caught by hauling a net; (2) the complete operation from setting or shooting a net to picking up or hauling. Also called: drag; (3) to heave up a trawl or seine net. Also called: wrap
hauling in, the process of pulling in fishing nets, lines etc. Also called: boarding the net
hauling leg. *See* haul up line
hauling rope. *See* purse line
haul loop, in a purse seine it is the loop at the wing end
haul seine. *See* beach seine
haul up line, a line with one end fastened to the double bag becket and the other to a small becket in the centre of the head line of a trawl. By hauling on this line the double bag becket is raised above the water to be hooked by the fish tackle, when the catch of fish can be heaved inboard. Also called: hauling leg
head cork, the main marker cork at the corners of a herring drift net
head, the tail end, trapping portion of a pound net. *Reference Fig. 32*
heading, (1) in a herring type drift net, the strengthening strip, five or more meshes deep at the top edge. Also called: board, guarding, hoody. *See* selvedge, also appendix (i); (2) the process of attaching double or strong netting to top and bottom edges of a sheet of netting. *See* selvedge

head iron. *See* shoe
head leak. *See* heading
headline, the rope on which floats are fastened. It is attached to the upper lip or top edge of a net. Also called: cork line, cork rope, float line, flue, head rope, net rope, top back, top line, top rope, upper taut. *Reference Fig. 19 & 23*
headline becket. *See* becket
headline leg, in a trawl, the wire leg between the dan leno and the headline. *Reference Fig. 23*
headrope. *See* headline
heart, the heart shaped portion of a fish trap, consisting of two wings designed to deflect fish into the trapping chamber. Also called: forebay
hedging. *See* leader
herring trawl, otter trawl, fitted with kites which ride above the headline. Having a longer cod end than usual, it is used for catching herring when on the sea bed. *See* appendix (c)
herring warp, a heavy hard laid rope, connecting a fleet of drift nets. Used for hauling and shooting. Also called: bush rope, fleet rope, messenger. *See* appendix (1)
hides. *See* chafer
hoody, (1) *See* selvedge; (2) *See* heading
hook, a section of netting which forms a pocket at the leading edge of a salmon bag net. Also called: spur. *See* pound net. *Reference Fig. 32*
hook rope, ropes which connect to quarter ropes to facilitate hauling
hoop net, (1) *See* fyke net 2; (2) *See* bully net
hula skirt. *See* chafer
hurdy gurdy. *See* gurdy
hydroplane float, improved design of float for increased opening height of trawl mouth

(I)

ice jigger, apparatus used to facilitate the setting of gill nets beneath an ice surface
independent piece, a length of wire between the towing rope and the bridle to facilitate disconnecting of the otter board. Also called: pennant. *Reference Fig. 23*
iron runner. *See* shoe

(J)

jackpole. *See* Bonito style fishing
Jap pole. *See* Bonito style fishing
jibs. *See* bats
jigger, plummet with hooks
jigger pole, a light pole used in trolling, it is set into the main outrigger troll pole at a slight angle. A jigger is rigged from its tip. Also called: gaff pole, spring pole, sucker pole
jigging. *See* trolling

jumper net, a small single staked trap net, set close into the shore. *See* pound net. *Reference Fig. 34 & 35*

(K)

kayacas. *See* Bohol trap net
keevil. *See* meshstick
keg, a small cask or barrel. With flagpole attached it is used for marking
keg line. *See* dan line
kelly's eye. *See* Figure of 8 link. The kelly's eye is shackled in to the back strop of an otter board, and effects jamming of aforementioned line. Also called: V.D. ring. *Reference Fig. 24*
kessock net, a very small herring drift net
kibble. *See* meshstick
kite, a small shearing device to accentuate the vertical fishing range of otter trawls. Used mainly for herring trawling. *See* appendix (c)
knotless net, when the meshes are formed with twisted yarn which is knitted together and not weaved with knots. Can only be made with a machine. Such netting has to be repaired with knots

(L)

lace hoods, the line along which the top and bottom parts of a trawl are laced together. Also called: lastridge
lacing, the join of two sections of netting by winding twine and fastening at intervals with a hitch
lam net, a net into which the fish is frightened by thrashing the water, shouting, and striking the boat with a stick, etc. Also called: splash net
Lampara purse seine, a surround net with the sections of netting made and joined to create bagging. Hauled with purse rings. Used for mackerel, pilchard, sardine, etc. *See* purse seine. *Reference* Chapter 9 *Fig. 39*
landing bag. *See* bag
lanyard. *See* gangen
large line fishing. *See* great lining
Larsen trawl. *See* floating trawl
lash, the rope which connects adjacent drift nets to the herring warp. Also called: strop, stopper
lastridge, (1) *See* lace hoods; **(2)** *See* belly line
laying out, (1) the proceedure of laying long lines. Also called: setting, shooting, stringing; **(2)** putting a net into the water for fishing purposes. Also called: setting, shooting, paying away
lazy decky, rope attached between the headline of a trawl and the cod end, used to facilitate drawing the cod end toward the vessel. Also called: lazy line, poke line, poop line, pork line
lazy line. *See* lazy decky

lazy rope. *See* double bag becket
lead. *See* leader
leader, in a pound net, it is the fence like barrier, usually of netting, which hangs vertically in the water, and guides fish towards the trap. Also called: fence, hedging, lead. *Reference Fig. 32*
lead line, (1) *See* ground rope; **(2)** the lower weighted rope of any encircling net or gill net
lead line strip. *See* selvedge
leads, the lead weights strung on the bottom of the lead line of a net. Also called: weights or sinkers
lead strip. *See* selvedge
leech line. *See* quarter rope
leg, (1) *See* bar; **(2)** a connecting length of rope, with an eye splice at each end
length, applied to netting, it means the stretched dimension of the clean meshes. *Reference Fig. 2*
lengthener. *See* extension piece
lever, the first row of half meshes in a net
lever net, fished from a raft. *See* lift net
lift net, a sheet of netting surrounded by a frame which effects the capture of fish in its tract as the net is lifted vertically. Also called: lever net. *See* appendix (j)
ligkop. *See* reef drag seine
line hauler. *See* gurdy
line lure fishing. *See* trolling
line roller. *See* gurdy
line trawl. *See* set line
lining, fishing by means of hand or long lines with baited hooks. Also called: tub trawling
link, a metal ring
lint, (1) a herring type drift net, particularly the body of the net. *See* appendix (i); **(2)** the small mesh netting between the two wall sections in a trammel net. *See* appendix (f)
live bait fishing. *See* Bonito style fishing
live box, receptacle in which crayfish etc., may be kept alive. Also called: crate, float, lobster car
lobster basket. *See* crayfish trap
lobster car. *See* live box
lobster pot. *See* crayfish trap
lobster trap. *See* crayfish trap
long-haul seine, a comparatively short beach seine, which covers a large catching area because of the distance hauled
long-haul seining, a method of fishing a seine with two boats, which head towards the shore where the net is landed by hand
long line, (1) a gear consisting of a main line supporting a large number of baited hooks, carried on short lighter vertical lines. Also called: box line, drift line, multiple-hook line, set line, trawl line, trot line, tub gear; **(2)** *See* tip line; **(3)** *See* great line

lower bag, in purse seine or other surround nets the heavier section where the fish are finally held for brailing, may be divided into two sections, i.e. a main heavy top bag and below a lower bag of slightly lighter material. *Reference Fig. 37*

lower wing, the part of a trawl net which is joined between the belly and the upper wing. The ground rope is fastened to the lower wing and the lower bottom meshes. Also called: bottom wing. *Reference Fig. 12 & 22*

lug, a short end of rope or a small length of rope connecting two main ropes

(M)

mackerel pocket, a large bag of coarse netting attached to the side of a mackerel seiner. It is held in position by a stand off stick. Also called: spiller

madrague, a trap used in the Mediterranean to catch tuna. Also called: pig catcher, tuna trap

main line, part of a long line to which gangen are attached. Also called: ground line

main outrigger troll pole, one of a pair of wooden poles, rigged on each side of a boat, for spreading the fishing lines beamwards. Also called: outrigger pole, outrigger troll pole

making a set, circling a school of fish while laying out the net

marker, (1) graduation on a trawl warp, a sounding line, or troll line; **(2)** *See* dan

marker lead, an identification lead, fitted at the corner of a herring drift net or ring net. *See* appendix (i)

mash. *See* mesh

mask. *See* mesh

mechanical harvester, mechanical equipment for the continuous catching of clams, scallops, etc. Also called: scooper

mesh, (1) two loops of twine, which are joined together with knots to form a diamond. Also called: mash, mask, web. *Reference Fig. 1*
(2) in net mending etc., it is one clean mesh. Also called pick up. *Reference Fig. 14*

meshes, the spaces formed when, following in continuity, one row of loops is joined with knots to the preceding row of loops

meshing, when fish are held fast in the meshes

meshing needle. *See* needle

meshpiece, an extension to the cod end of a trawl

meshpin. *See* meshstick

mesh size, the dimension of the mesh, from the centre of one knot to the centre of the next diagonally opposite knot. *See* stretched mesh. *Reference Fig. 1*

meshstick, a gauge which is used to regulate mesh size when making nets. Also called: bowel gauge, keevil, kibble, meshpin, moot, pin, shale, spool

messenger, (1) in an otter trawl, a wire rope with a manila tail for heaving the warps together into the towing block after the gear has been shot; **(2)** *See* herring warp

messenger hook, a specially shaped hook attached to messenger

mid-water trawl. *See* floating trawl

molgogger, a roller used on drifters when shooting the nets. Also called boggey, cage roller

moot. *See* meshstick

mounting. *See* hanging 2

mouth, entrance of a net

multiple hand line, a single vertical hand line with a series of barbed hooks attached to it by 'spreaders' at regular intervals

multiple-hook line. *See* long line 1

(N)

narrowing. *See* bating

needle, a pointed instrument made of wood, aluminium or plastic, in which the twine is wound for the purpose of making nets by hand. Also called: meshing needle, netmaker's needle, netting shuttle

net, fabric knotted into meshes by continuously joining one row of loops to another. The knots used may be made with various ties, single or double. Also called: webbing

net depth, (1) when relating to a completed net of specific design it is the distance from the headline to the ground rope, usually designated by the number of meshes; **(2)** with a machined sheet of netting, it can mean the number of cut meshes along one edge of the sheet; **(3)** with hand made netting it may indicate the number of clean meshes which have been made

netmaker's needle. *See* needle

net rope. *See* headline

netting. *See* net

netting shuttle. *See* needle

Newfoundland cod trap, movable box like trap, with a leader. It is placed in inshore water

nigger head. *See* gipsy

non-return valve, an arrangement of netting attached on the inner side of a fish net. It tapers towards the holding part of the net, thus preventing the escape of fish—for instance—in a drag net, seine or trawl, it will be cone shaped and may be called a flapper, funnel, or pocket. *Reference Fig. 22*

Similarly in crayfish traps (lobster pots etc.) the non-return valve is cone shaped and may be called an eye or funnel. *See* appendix (g)

In the larger type of trap nets there may be one or two pairs of vertical sheets of netting, set at an angle, they may taper very slightly from the sides to the centre of the trap. These may be called scales or walls. *Reference Fig. 32*

no overhang trawl, type of shrimp trawl with identical top and bottom sections

norsals, (1) short lines spaced regularly along the top edge of a net (especially a drift net) for connection to the headline or lead line, to give the netting a looseness. Also called: daffins, norsells, nossles, ossles. *Reference Fig. 17*
(2) a 'gangen' as used in long lining, may also be called a norsal etc., and vice versa. *See* appendix (q)

norsalling, the procedure of attaching short lines between the top or bottom ropes, and the appropriate reinforced edge of a net. The norsals are for instance, spaced every five or six meshes on a herring drift net. This gives the net a certain looseness

norsells. *See* norsals
nossles. *See* norsals

(O)

ossels. *See* norsals
osselling. *See* norsalling
otter, fishing tackle consisting of a short weighted plank, fitted with fish lures; controlled from the shore it tends to shear outwards when pulled through the water
otter board, one of a pair of shearing devices, which keep the mouth of trawl horizontally open during towing. Made of wood, strengthened and weighted heavily, its shape is usually rectangular. When fitted with brackets and other accessories its design is such that it will shear outwards when pulled through the water. Also called: board, door, sheer board, spreader, trawl board, trawl door. *Reference Fig. 24*
otter board bracket. *See* bracket
otter door. *See* otter board
otter trawl, a cone shaped drag net which is pulled by a vessel over the sea floor. It has various attachments for keeping the mouth open during towing, such as floats, weights, and principally two otter boards, which shear outwards away from one another and thus spead the net. *Reference Fig. 25*
outrigger, when a second pound net is set directly on the end of another such net

outrigger pole, (1) *See* stand off stick; **(2)** *See* main outrigger troll pole
outrigger troll pole, *See* main outrigger troll pole
overhang. *See* cutback
overhang trawl. when a trawl net is shaped and rigged to allow the headline to fish ahead of the ground rope. The difference may be referred to as the overhang
oyster dredge. *See* dredge
oyster tongs, a pair of rakes with a fulcrum used to gather oysters

(P)

Pacific trawl. *See* box type otter trawl
pair net. *See* paranzella fishing
pallet. *See* bowl 1
paranza. *See* paranzella fishing
paranzella fishing, an old and still used method of pair trawling—that is towing a bottom trawl between two boats, which fish approximately half a mile apart. Consequently the net is automatically spread and no otter boards are required

The net is usually larger than the normal deep sea otter trawl, though not as high, and the wings are shorter. Also called: pair net, paranza, pareja, Spanish trawl, tartanas, twin trawl
pareja. *See* paranzella fishing
parmenta, a lampara purse seine, used for catching green turtle
paying away. *See* laying out
pegging, hand lining with a sinker and hook attached. The line is jerked upwards a fathom or so as it strikes the bottom
pelagic trawl. *See* floating trawl
pellet. *See* bowl 1
pennant, (1) *See* independent piece; **(2)** wire which connects the junction where the bridle meets the towing rope. The wire used for similar connections
percentage line, *See* whiskey line
Peter net, a salmon set net with floats along headline and weights along the lead line. One end is fastened to shore, and the other anchored in the sea, at right angles to the coast line
phantom trawl. *See* floating trawl
pick, instrument used to dig clams
picking, harvesting of oysters by hand when the tide is out
pick-up, (1) *See* take up; **(2)** *See* mesh
piece. *See* basket
pig-catcher. *See* madrague
pin, (1) *See* meshstick; **(2)** a fastening which is driven or sunk into the beach for holding the rope from a pound net stake. *Reference Fig. 35*

pinch off, isolating a section of the catch of a purse seine, for immediate brailing. Alternate to zipper

plug, an artificial lure resembling a small fish

pocket, (1) *See* cod end; **(2)** *See* non-return valve; **(3)** *See* fish court

point, one cut mesh. Also called: side knot, diamond. *Reference Fig. 14*

poke line. *See* lazy decky

pole line. *See* tag line

pole net. *See* beam trawl

pony board. *See* dan leno board

pole trawl. *See* beam trawl

poop line. *See* lazy decky

pork line. *See* lazy decky

porpoise cover, a protective cover of heavy mesh netting, fitted to a trawl net

pot, (1) *See* fish court; **(2)** *See* crayfish trap

pot warp, a rope connecting a crayfish trap on the ocean bottom with the buoy floating on the surface

pound. *See* fish court

pound head. *See* head

pound net, a set net. The trap portion is composed of netting with vertical sides, a top, a cover, and non-return valves fitted inside. This may be moored with anchors and casks and held open with stretcher poles, also floats. *See* non-return valve. *Reference Fig. 31, 32, 33, 34 & 35*

Again the trap may be secured to stakes embedded in a beach. All the edges of the netting are strengthened with ropes.

Usually a long sheet of netting which may be fitted with floats, or staked, is fastened near the mouth of the trap and laid in towards the shore.

Pound nets are often set in patterns, for instance, one net may be fixed directly on the end of another. Also called: fish trap, fly net, jumper net, trap net, bag net

power gurdy. *See* gurdy

preventer chain, a chain which is passed round the towing ropes when trawling in bad weather conditions. It is fastened to deck fittings and is a safety precaution. Also called: stopper chain

pull net. *See* drag net

puppy, the first net in a fleet of drift nets, that is the one which is closest to the drifter. Also called: stem net

purse, (pursing) the operation of drawing in the purse line of a purse seine net to complete the enclosure. *Reference Fig. 40*

purse lampara. *See* lampara purse seine

purse line, rope used to close the bottom of a purse seine. Also called: hauling rope, purse rope, purse string, pursing cable. *Reference Fig. 40*

purse line rings, metal rings joined at intervals with rope bridles to the bottom of a purse seine to hold and guide the purse line. *See* purse line. *Reference Fig. 37*

purse ring slip knot, a simple device for releasing the last six or nine wing end purse rings from the purse seine

purse rope. *See* purse line

purse seine net, a net, rigged with floats and weights, which is cast in a circle around a school of fish. The wings are usually tapered. It is fitted with a closing device, consisting of a purse line, which passes through metal rings attached, at intervals, to the foot of the net with short rope bridles. When this rope is hove inboard, the bottom of the net automatically closes and the surplus netting is pulled into the vessel, thus creating an artifical pond. The entrapped fish can then be scooped out with brail nets. *Reference Chapter 9 Fig. 40*

In common with other main fisheries, the designs of these nets, as well as the pursuing operations, do vary. For example, a net may be quite small or again extremely large. The wings may taper slowly or be shaped steeply. This may be achieved by joining together rectangular panels of netting which decrease in size, or by actually cutting the taper. The method of operation usually determines the position of the landing piece, whether it is at the end or in the centre of the net. *See* bag

The laying of the net may be carried out by a skiff working in conjunction with a parent boat which does the actual hauling. Alternatively one vessel only may complete the whole operation. Also called: round haul net. *See* ring net

purse string. *See* purse line

pursing cable. *See* purse line

push net, triangularly framed shallow mesh bag operated by hand. It is fished in shallow water by a horizontal pushing motion

(Q)

quarter line. *See* quarter rope

quarter rope, rope fitted around the outside of a trawl net, between the rollers and the dan leno. Two such ropes are used to facilitate the hauling of an otter trawl. Also called: leech line, quarter line. *Reference Fig. 23*

quarters, the two reinforced junctions of double netting, where the top wings meet the square. *Reference Fig. 12*

quarter strop, one of two wire strops used in trawling, which, when fitted with swivels, join between the quarter ropes and the rollers

(R)

rake, (1) *See* dredge; (2) *See* digger

reef drag seine, a net similar to a beach seine, which is dragged over rough ground or reefs, with the headline submerged and the ground rope continuously freed of snags by divers. Also called: ligkop or reef seine

reef net, set net held in place by four anchors and worked with two boats

reef seine, *See* reef drag seine

reining, the strengthening strip of netting five or more meshes deep at the foot of a herring type drift net. Also called: guarding. *See* selvedge

rib line. *See* belly line

rigging, the process of fitting the necessary ropes and accessories so as to make a net ready for fishing. *See* hanging. *Reference* Chapter 5

ring net, (1) a net comprising several rectangular sections of netting, which, when joined together form a large wall. This is mounted to ropes, and fitted with floats and weights.

The mesh size usually decreases and the twine size alters towards the centre of the net, where the fish are entrapped and held for brailing.

During the fishing operation, the net is laid in the water in a circular motion, in the process of surrounding a shoal. A hawser attached with rope stoppers to the foot of the net, is hauled, pulling the bottom of the net towards the vessel. The wings are also hauled into the boat, making a pond where the fish are held for brailing.

Two boats are required for the operation. In shallow water the ring net may hang from the surface to the sea floor. *Reference* Chapter 9 *Fig. 36 & 37*
(2) *See* purse seine net

ripe line. *See* belly line

ripper. *See* squid jig

roller, roller or bobbin, made of wood, iron, rubber or other hard wearing material, they are designed to roll over the sea bed. Fastened to the ground rope of a trawl, they safeguard the lower lip of the net when fishing over rocks or rough ground. They may be of various shapes, principally, round, disc or spherical. Also called: bead bobbin, disc roller, setting bobbin. *Reference Fig. 23*

rolling bobbin. *See* roller

rough meshes, the meshes with the direction of the knots on a sheet of netting. *See* cut meshes

rough net, a type of gill net

round, (1) a continuous line of half meshes, or the distance between one line of knots and the next; (2) the initial line of loops when beginning to hand braid a net

round haul net. *See* purse seine net

run, the direction in which the knots are made, on a sheet of netting. Also called 'scone'. *Reference Fig. 2*

runner. *See* shoe

running tuck, a method of working a stop seine in a strong current, so as to bring the wings of the net together

(S)

sablefish line, a long line consisting of 10 to 12 baskets with 200 to 225 hooks per basket

sack. *See* bag

salmon bag net. *See* pound net

salmon trap. *See* pound net

sardinal net, a very long sardine drift net

scale, *See* non-return valve

scare, an assembly used on a purse seine net to prevent the escape of fish while the two ends of the net are being closed during pursing

scare line, a device used to scare fish in the direction of the fishing operation. *See* Bohol trap net

schooling up, grouping a large number of fish near the surface

scone. *See* run

scoop. *See* dip net

scooper, a type of dredge, mounted on runners and used to catch oysters

scooper type of harvester. *See* mechanical harvester

scoop net. *See* dip net

scoop seine, a small type of seine which is used as an accessory gear for taking the catch from the large semicircular enclosure of deep-water fish barricades, which are devoid of a collecting chamber

score, in relation to the depth of a net it indicates 20 meshes, i.e. a drift net of 360 meshes deep would be referred to as an 18 score net

scraper. *See* dredge

scull, a long flat basket for holding small lines

scum net. *See* thief net

sea horse, a hoe with long prongs and handle, used for collecting clams

sean. *See* seine net

seine net, a net with a small conical bag and long narrow wings. It is rigged with floats and weights. When fishing a triangular pattern is made, laying first several coils of rope, dependent on conditions, then the net, and finally more rope. Both ropes are hauled, and any fish in the area enclosed are frightened into the bag. Also called: deep sea seine, mid water seine, sean, seyne or snurrevaad. *Reference Fig. 13, 19 & 20. Also* appendix (o)

selvage *See* selvedge

selvedge, this reference has such a wide range of application, although in every case it refers one way or another to the edges of netting. Examples follow: **(1)** It can mean any of the four edges round a sheet of netting; **(2)** in hand braiding, it refers to the side edges only; **(3)** it may indicate a strengthening strip of netting attached to the edges of a plain sheet of netting during manufacture. For instance (a) a double mesh selvedge all round suggests a sheet of netting requires edging around with a mesh made of double twine, (b) a heavy row selvedge top and bottom, would mean making the net with one row at the top and bottom of the net with a heavier twine.

A strip of heavy netting several meshes deep is used for strengthening drift nets or surround nets. Also called: guarding, heading

semi-purse net, a light purse seine net. *See* purse seine net

set, **(1)** when a fishing assembly is temporarily secured on the sea bed, usually with anchors and weights, so that it will not move. *See* appendix (e); **(2)** *See* laying out

setback. *See* cutback

set gill net, **(1)** *See* set; **(2)** *See* gill net

set impounding net, **(1)** *See* set; **(2)** *See* impounding net

set line, **(1)** *See* set; **(2)** a long line which is secured on the sea-bed with moorings. Also called: bottom line, ground line, line trawl, set long line, trawling line, trawl line

set long line. *See* set line

set net. *See* set

set of gear, the entire net assembly of a trawl

setting. *See* laying out

setting bobbin. *See* roller

seyne. *See* seine net

shack fishing, method whereby men fish from the deck of the boat each with a number of lines, and being responsible for his own gear and catch

shale. *See* meshstick

shank net. *See* dredge net

shark cover, a protective cover of heavy mesh netting fitted to a trawl net

shark gill net. *See* gill net

sharks mouth. *See* cutback

sheer board. *See* otter board

shipping drum. *See* gipsy

shock, a fitting of elastic material on a trolling line, which absorbs part of the strain on the assembly when a heavy fish strikes the lure. Also called: shock absorber

shock absorber. *See* shock

shock line. *See* tag line

shoe, a metal framework used in beam trawling. Fitted to each end of the beam they activate a sledging motion over the sea-bed, simultaneously holding the beam up to form the mouth. Also called: head iron, iron runner, trawler head, trawl head, trawl iron. *See* appendix (b)

shooting. *See* laying out

shooting in piece, a rectangular piece of repair netting for a trawl net

shore seine. *See* beach seine

shore weir. *See* barricade

shoulder, **(1)** the mid netting sections between the bag and wings in a ring net, purse seine, beach seine and demersal seine. *Reference Fig. 19, 20 & 37*

NOTE, the shoulders and wings of a purse seine may be called the body or bunt of the seine

shrimp trawl, small type of beam or otter trawl

shrinking. *See* bating

side, the vertical outside section of a pound net

side cord, a light cord joined along the top and lower edges of a drift net

side knot. *See* point

single anchor line. *See* drop line 1

single bag becket, a strong becket, to facilitate heaving a very heavy or fouled cod end

single hand line, a single vertical hand line with one or more hooks. Also called: drop line

sinker, **(1)** a weight used to sink a fishing line; **(2)** several sinkers may be spaced along the ground rope of a fishing net

sink gill net, a gill net weighted sufficiently to sink the headline so that the net fishes close to the bottom

skagen net, high opening type of trawl which may be fished using the seine net principle or as a conventional trawl. Also called: wing trawl. *Reference Fig. 27*

skate, a square of tarpaulin used in the halibut fisheries as the receptacle for a unit of long lines

skirt, a term used when several meshes at the foot of a gill net or seine are made of heavier twine than the body of the net

sleeve net, cylindrical netting open at both ends

sling, **(1)** in a purse seine, the upper end of the breast purse line, which is fastened to the base of the davit loop; **(2)** the top of a ring net; **(3)** a large repair piece for a ring net bag, i.e. 400 meshes × 60 yd.

sling-ding, in hand-line fisheries, it is a light galvanized iron rod, with an eye at each end, from which gangens are hung

small line, Also called: spiller. *See* haddock line

snap net, a net with hinged or folding wings, which can be snapped closed when fish enter

snare, a slip noose made with twine and attached to the end of a rod

snell. *See* gangen

snood, (1) *See* gangen; **(2)** *See* norsals; **(3)** any one of the side ropes of a pound net between the top and the lower parts of the frame. *Reference Fig. 32*

snoot line. *See* gangen

snurrevaad. *See* Danish seine

soger, a hardwood stopper, fitted at intervals along the vessels rail to prevent hand lines from slipping

sole rope. *See* ground rope

Spanish trawl. *See* paranzella fishing

spiller, (1) one of the components of a pound net; **(2)** *See* mackerel pocket; **(3)** small line

spindle, metal bar which slips through dan leno bobbin (or roller)

splash net. *See* lam net

splitting strop. *See* double bag becket

spool. *See* meshstick

spoon, a curved piece of metal with a hook attached, which is used in trawling for attracting fish

spoon bait. *See* spoon

sprawl wire. *See* spreader 1

spreader, (1) a device used in hand line fishing to spread gangens so that they do not foul one another. Also called: sprawl wire; **(2)** *See* dan leno; **(3)** *See* stretcher 2; **(4)** *See* otter board

spreader bar, a strong metal bar used in trolling. It has an eye at each end, and is fitted at the junction of the tag line, main line and in-haul

spring line. *See* jigger pole

spring rope. *See* purse line

spur. *See* hook

square, the section of netting in a trawl net which overhangs the ground rope. It is fitted between the top body and the two top wings. *Reference Fig. 12 & 22*

squaw net, a small gill net, fastened to a pole, which in turn is secured to another pole on the shore

squid, an artificial lure

squid jig, an arrangement of hooks for catching squid and the like. Also called: ripper, squid jigger

squid jigger. *See* squid jig

squid method. *See* bonito style fishing

stab net, a bottom type gill net which is fished in deep water

staff, a pole to which gill nets are attached

stangeon. *See* gangen

stake, one of a number of long wooden poles, to which certain nets are secured, i.e. A pound net *Reference Fig. 35*

stake net, (1) a gill net which is hung on stakes. It is covered at high tide and the fish are trapped as the water recedes; **(2)** a pound net which is held up with stakes; **(3)** a weir which is supported with stakes

standing line. *See* tag line

stand-off stick, a hinged pole, fitted to the rail of a seiner. It prevents the net pulling the skiff alongside the vessel because of a heavy catch, thus leaving sufficient space for brailing

stapling, the spacings or loops formed when hanging netting to a line, by lifting a number of meshes and hitching at intervals. Also called: bulls

starter three-legger. *See* halfer

stealing. *See* bating

steel warp. *See* towing rope

stem net. *See* puppy net

stock brail, a power driven dip net for brailing the catch from a purse seine

stole meshes. *See* bating

stolers. *See* bating

stomach piece. *See* apron

stop link. *See* Figure of eight link

stop net, a net which is used for blocking the mouth of a bay etc.

stopper, (1) *See* lash; **(2)** a rope used for attaching the bowls on a fleet of drift nets; **(3)** seizings around a main line which prevent the leader stops from slipping along the main line

stopper chain. *See* preventor chain

stop seine, (1) a short seine which is laid across the wings of a larger seine, thus enclosing the fish in an artifical pond; **(2)** a small seine which is used to collect the fish caught inside a large surrounding net. Also called: tuck seine

stosh. *See* chum

stowing. *See* bating

stow net, a large funnel or cone shaped net which is lowered beneath an anchored boat or raft, to catch shoals of sprats as the fish come up the estuary on tide. *See* boom net. *See* appendix (d)

straight selvedge. *See* double selvedge

strainer, a temporary secondary bag of larger mesh fitted to a purse seine net for allowing the small fish to escape

stream net, stake net fished in fast running water, fitted with intermediate sticks which cause bagging where the fish are trapped as the tide recedes

stretch measure, the longer measurement along the side of a sheet of netting when all the meshes are closed. *See* stretched mesh. Also called: stretched measure. *Reference Fig. 1 & 16*

stretched measure. *See* stretch measure

stretcher, (1) *See* dan leno; **(2)** wooden poles for spreading certain parts of a pound net. *Reference Fig. 33 & 34*

striker method. *See* Bonito style method

string, part of a complete long line. It is usually 60 fathoms long

stringing. *See* laying out

strop, (1) *See* lash; **(2)** a wire rope becket

submarine trap, a trap net which is set in deep water. Also called: deep water trap

sucker pole. *See* jigger pole

Sueberkrueb trawl door, narrow type of otter board used for pelagic trawling. *Reference Fig. 29*

sunk net, a drift with the warp rigged over the net

sweep, (1) the dimension of the aperature at the mouth of a net; **(2)** *See* bridle 1; **(3)** the area covered between the mouth of a trawl net and the two opposing otter boards

sweep line. *See* bridle 1

sweep net. *See* beach seine

swivel, a steel connecting device which turns within itself, thus, when shackled to ropes, turns and kinks are avoided

swum net, when the herring warp is fitted beneath a drift net

(T)

tag line, used in trolling it is the line from the main outrigger troll pole to the main line at its junction with the in-haul line. Also called: brail line, drag line, pole line, shock line, standing line

tail, *See* cod end

take up, (1) the percentage of netting distributed along a rope when hanging in the net. *Reference Chapter 5*

(2) when joining two sections of netting of different sizes, it is the percentage of netting which has to be distributed so that they join evenly. Also called: pick up

take up mesh. *See* bating

taking in. *See* bating

tan. *See* barking

tangle net, flimsy net with the netting hanging loosely or in a specific way so that fish striking it become immediately entangled

tannage. *See* barking

tanning bark. *See* barking

tarring, the preservation of nets by immersing them in tar

tartanas. *See* paranzella fishing

the soak, the period during which a line is left fishing on the sea bed

the stick. *See* dan leno

thief net, a secondary net used for catching herring which fall out of a drift net as it is hauled inboard. Also called: scum net

three legger. *See* halfer

throat, the narrowest section of the body of a trawl net, nearest the cod end. More commonly known in herring trawling, when an extension is fitted on to the bellies of the conventional otter trawl. This extension which is long and narrow is referred to as the throat or funnel

throat line, a long line, used for catching cod. Also called: boulter, buffow line, bultow, thrott line, trawl

thrott line. *See* throat line

throw net. *See* cast net

tickler, a chain, positioned across the mouth of a drag net when fishing for flat fish. Its purpose is to churn up the bottom

tickler chain. *See* tickler

tide net. *See* stake net 1

tie line. *See* trip line

tip line, used in trolling it is a short and lighter line, fixed at the far end of the main line. Also called: tippet, trolling leader

tippet. *See* tip line

tom. *See* gangen

top back. *See* headline

top bag. *See* bag 3

top line. *See* headline

top rope. *See* headline

top wing. *See* upper wing

towing, dragging a fishing gear along the sea floor

towing block, a collapsible metal framework, on which the towing rope rests after the trawl has been shot. Fitted near the stern its purpose is to hold the ropes clear of the propeller. *See* appendix (m)

towing rope, extremely long cables of rope or steel which are used to tow trawl nets. Also called: drag rope, towing warp, trawling cable, trawl warp, warp

towing warp. *See* towing rope

tow leg, in trawling it is a rope which connects the foot of the dan leno with the ground rope end of the wing. *Reference Fig. 23*

trail. *See* dan leno

trailing. *See* trawling

trail net, the furthermost net in a fleet of drift nets

trammel net, a net made with three sheets of netting which are joined so that they lay together in one wall. The two outer sides are of a larger mesh than the loosely hung inner piece. It is fitted with ropes, floats and weights and set on the bottom. When a fish strikes the net it becomes entangled in a pocket, as the force carries the smaller mesh netting through the larger mesh. *See* appendix (f)

trap net, (1) *See* barricade; **(2)** *See* pound net; **(3)** *See* crayfish trap

trawl, (1) a conically shaped bag of netting, with the mouth designed in numerous ways to suit the different devices for maintaining the opening whilst being towed through the water; **(2)** *See* throat line

trawl beam. *See* beam trawl

trawl board. *See* otter board

trawl bridle. *See* bridle

trawl door. *See* otter board

trawler head. *See* shoe

trawl gallow. *See* gallow

trawl head. *See* shoe

trawling, fishing with a trawl. Also called: dragging, tracking, trailing, trawl fishing. *See* appendix (m)

trawling cable. *See* towing rope

trawling line. *See* set line

trawl iron. *See* shoe

trawl line, (1) *See* set line; **(2)** *See* long line

trawl lining. *See* long line

trawl net. *See* trawl

trawl plane, a device fitted to the headline for increased opening height. *Reference Fig. 30*

trawl toad, a device used in mid-water trawling

trawl warp. *See* towing rope

trawl warp mark. *See* marker

trawl winch, the mechanism used for shooting and hauling a trawl. *See* appendix (m)

triangular bracket. *See* bracket

trident, a three-pronged fish spear

trim net, a bag net used for catching white bait and eels. It is smaller than a stow net but similar. *See* stow net

trip line, a light rope threaded through metal rings at the tail end of a net. Used for closing and releasing. Also called: tie line

tripper, a float with rope tail attached to anchor of a trap net to prevent loss

trolling, fishing with the use of lured long lines which are dragged through the water. Also called: jigger, line-lure fishing

trolling leader. *See* tip line

trolling line. *See* troll line

trolling spoon. *See* spoon

trolling wire. *See* troll line

troll line, a line with hooks and lure (live or artificial) which is towed behind a moving boat

trot line. *See* long line

trow net. *See* cast net

tub. *See* basket

tub gear. *See* long line

tub trawling. *See* lining

tucking, clearing a larger net with the aid of a tuck net. *See* stop seine 2

tuck net. *See* stop seine 2

tuck seine. *See* stop seine

tuna bait fishing. *See* Bonito style fishing

tuna trap. *See* madrague

tunnel, (1) *See* funnel; **(2)** *See* non-return valve

twin trawl. *See* paranzella fishing

(U)

undercut. *See* cutback

unit. *See* basket

up and down line, (1) *See* breast purse line; **(2)** *See* drop line

upper taut. *See* headline

upper wing, a section of a trawl which is laced between the square and the lower wing. *Reference Fig. 12 & 22*

(V)

V.D. ring. *See* Kelly's eye

V.D. trawl. *See* Vigneron Dahl trawl

vertical line. *See* gangen

Vigneron Dahl trawl, an otter trawl fitted with Vigneron Dahl bridles and the otter boards attached some distance from the tip of the wings, thereby increasing the effective fishing width of the net. *Reference Fig. 23*

(W)

wall. *See* trammel net. It is one of the outer panels of netting

wall of netting, vertical piece of netting

warp, (1) *See* towing rope; **(2)** *See* herring warp

warping end. *See* gipsy

weaving. *See* braiding

web. *See* mesh

webbing. *See* net

weight, (1) *See* lead; **(2)** *See* sinker

weir, (1) *See* barricade; **(2)** *See* pound net

Western trawl. *See* box type otter trawl

whammel, a salmon drift net

whiffing line, a hand line

whipping drum. *See* gipsy

whiskey line, a long fishing line, used in trolling. Made fast to the mast or a boom it fishes further out than the other stern lines. Also called: dog line, percentage line

white sea bass purse seine. *See* purse seine net

windlass, small heaving mechanism

winch. *See* trawl winch

winch end. *See* gipsy

winch head. *See* gipsy

wing, (1) the section of netting at the end of a ring net, purse seine net, beach seine net, etc. *Reference Fig. 37*

(2) the forward section of netting (or arm) of a demersal seine net, of which two are necessary to guide fish into the bag. *Reference Fig. 19 & 20*

(3) one of the four wings required to make up a trawl, i.e. Two top wings and two lower wings. The wings direct the fish into the bag. *Reference Fig. 22*

wing board, a shearing device used in mid-water trawling. Works on a similar principle to an otter board

wing end, (1) the end of a purse seine which is hauled first, after laying the net **(2)** the end of a trawl or seine net wing. *Reference Fig. 12*

wing lines, the ropes connecting the two ends of the headline to the two ends of the ground rope

wing trawl. *See* skagen net

wrap. *See* haul

(Y)

yard seine. *See* beach seine

(Z)

zipper, opening edge along the body of the net of a mid-water trawl, facilitates brailing

zipper lines, ropes which are fitted to large purse seines and can be used for sectioning large catches.

ISBN 0 85238 031 3

Price £1·75